To George Hunt

with best wishes

Hans Blokpoel
26/11/80

BIRD HAZARDS TO AIRCRAFT

BIRD HAZARDS TO AIRCRAFT

Problems and Prevention of Bird/Aircraft Collisions

H. BLOKPOEL

Published by Clarke, Irwin & Company Limited
in association with
the Canadian Wildlife Service,
Environment Canada and
the Publishing Centre, Supply and Services Canada.

Canadian Cataloguing in Publication Data

Blokpoel, H., 1938-
Bird hazards to aircraft

Bibliography
Includes index.
ISBN 0-7720-1086-2 bd. ISBN 0-7720-1087-0 pa.

1. Airports - Bird control. 2. Aeronautics - Accidents. 3. Bird control. I. Title. II. Canada. Wildlife Service.

TL725.3.B5B56 614.8'69 C76-017100-9

Govt. catalogue no. CW66-47/1976

Hardback ISBN 0-7720-1086-2
Paperback ISBN 0-7720-1087-0

Published simultaneously in the United States by Books Canada Inc., 33 East Tupper Street, Buffalo, N.Y. 14203, and in the United Kingdom by Books Canada Limited, 1 Bedford Road, London N2.

1 2 3 4 5 JD 80 79 78 77 76

Printed in Canada

This book is dedicated to
my mother and the memory of my father.

Acknowledgements

This book was written at the request and under the auspices of the Associate Committee on Bird Hazards to Aircraft of the National Research Council of Canada. Several members and ex-members, particularly W.H.S. Bird, A.J. Bosik, H. Boyd, H.R. Finley, G.M. Geekie, C.E. Hansen, R.M. Kidd, J.W. Noonan and A.B. Simpson, gave expert advice. The enthusiasm and comments of V.E.F. Solman, the Chairman of the Committee, were of great help in bringing the book to completion. M.S. Kuhring, who chaired the Committee from 1962 to 1973, provided technical material and helped compile bird strike statistics and Appendix 7-1.

Several other people contributed by giving advice, making comments and suggestions, or by providing certain information or references. Their names are: Th. Alerstam, M. Blokpoel, H.E. Bryant, T. Brough, A.P. de Jong, V.E. Ferry, H.S. Fowler, R.W. Fyfe, A.J.W. Hoorn, E.W. Houghton, F.R. Hunt, A.H. Joensen, J. Karlsson, W. Keil, M. Laty, G. Lid, M. Louette, J. Thorpe, J.M.M. van der Heyde, and L. van't Hof.

The competent editing of L.W. Billingsley increased the readability of the book while decreasing its bulk, and, hence its price.

Financial assistance was received from the Ministry of Transport and the Department of National Defence through the Associate Committee.

Table of Contents

List of figures

List of Tables

List of Appendices

Foreword

Birds and aircraft have collided with damage to both almost from the beginning of aviation. Early aircraft flew at such low speeds that birds had little difficulty getting out of the way. Nevertheless, the first recorded loss of a human life in an aircraft crash caused by a bird occurred in 1912. Since then, as both the speeds of aircraft and the number of aircraft flying have increased, so have incidents involving bird collisions.

When turbine power began to replace piston engines in the 1950's, the problem increased in importance. In 1960 the crash of a passenger airliner resulting from bird ingestion in turbine engines resulted in the loss of more than 60 human lives.

Military aircraft are exposed to a more serious risk because some exercises involve high speed at low altitude, where more birds are commonly present. Losses of military aircraft have been numerous and costly. In some countries lives have been lost as well. The loss of and damage to equipment continues to be of concern, but the loss of human life has not been great in recent years.

The difference between loss of human life and serious damage to aircraft without that loss depends on the circumstances of the bird strike. A difference of six inches in the point of impact can change the result from structural breakage to loss of life. Luck has played too large a part in the past for us to be complacent.

The problem is serious. Canada had to move into the field, study it, and apply successful results quickly. The Canadian bird hazard study and reduction program began earlier than some others and has been effective in reducing significantly the cost of replacing broken aircraft parts. There has been no loss of life on scheduled passenger carriers in Canada as a result of birds.

By applying the best information available, as presented in this book, we can reduce significantly the bird hazard. Those who work in the field know that the real solution to the problem depends on the knowledge and dedication of the people who apply corrective measures. High motivation, on a continuous basis, is the weapon that can most effectively reduce the incidence and danger of bird strikes.

This book was written at the request of the Associate Committee on Bird Hazards to Aircraft, and under its auspices. I hope that it will increase awareness of the problem and contribute to its solution.

Victor E.F. Solman,
Chairman,
Associate Committee on Bird Hazards to Aircraft,
National Research Council of Canada,
Ottawa, Ontario.

A suggestion to readers

If you are pressed for time, I suggest you scan the Table of Contents or the Index to find the topic that interests you most.

All chapters deal with different subjects and can be understood by themselves. Although the chapters are arranged in what seems to be a logical order, you need not follow this sequence.

H.B.

Introduction

Collisions between birds and aircraft, usually known as bird strikes, have occurred since the earliest days of aviation. The first recorded fatality due to a bird strike occurred in 1912, when a gull became entangled in the exposed control cables of an aircraft. During the first few decades of aviation, however, bird strikes happened infrequently, did relatively little damage, were rarely reported, and were not considered to be really serious.

In more recent times, with the introduction of new high-speed aircraft powered by vulnerable turbine engines, and with the phenomenal increase in air traffic, bird strikes have evolved from a minor nuisance into a serious and costly problem.

The bird strike problem has received a great deal of attention from the news media, in part because many people are interested in birds and in aircraft, but also because the layman can easily visualize the mechanics of birds hitting aircraft. Any motorist going at fifty miles an hour who has had the windshield cracked by a flying pebble will have little difficulty imagining the impact of a ten-pound goose on an aircraft which is travelling at 500 mph.

Much has been said and written about bird strikes, but as far as is known there is no comprehensive book on the bird hazard problem in general. This book attempts to deal with all aspects of the problem by (1) providing information about birds in general, relevant to the problem, (2) reviewing published bird strike statistics to indicate how widely they occur, (3) discussing present techniques to deal with the problem, (4) supplying a list of publications and other references, and (5) providing a list of names of individuals and organizations which the reader may consult for further information.

The book is primarily written as a reference book and guide for those directly involved in or concerned with bird strikes: pilots, air traffic controllers, airfield maintenance crews, airport managers, flight safety officers, aircraft designers, administrators, planners, biologists, and environmentalists.

Because people of different disciplines and backgrounds may use the book, and because a wide variety of topics is dealt with (varying from methods of reducing earthworm numbers to radar detection of birds), the basic principles underlying some of the techniques are explained briefly and examples are given where possible. In this way the book will also be of interest to the layman whose hobby is birds or aircraft.

Although biologists and aircraft engineers working on the

bird hazard problem may not find anything novel regarding their own specialty in this book, the designer of windshields may find the data on bird heights (and ambient temperatures at those heights) useful, and the biologist who is studying gull movements in the vicinity of airfields may be interested in the design of jet engines and their vulnerability to bird impact.

A harried airport manager with a particular bird problem is advised to scan the Table of Contents and to use the Index. It is unlikely that he will find a clearcut answer to his problem (otherwise this book need not have been written), but there is a good chance that he will find some relevant background information and references that will give him some useful clues as to what to do and to whom to go for help.

General information on birds and bird migration can be found in Chapter 1. Chapter 2 reviews some world-wide bird strike statistics. In Chapter 3 techniques for "bird-proofing" aircraft, i.e., making them more resistant to bird impact, are described. Attempts to develop on-board devices to disperse birds in front of the flying aircraft are reviewed in Chapter 4. Chapter 5 provides an account of the various methods to reduce bird numbers at airports, and Chapter 6 details the techniques to prevent bird strikes away from airports. Finally, Chapter 7 discusses the different organizations that work on the bird strike hazard. The chapters can be read in any order because each deals with a different subject and tells its own story.

The appendices to Chapters 1 through 7, the list of abbreviations, the references, the photo credits, and the index complete the book. The reference section lists the references numerically in alphabetical order.

Bird strikes are a world-wide problem. Data from many countries are included in this book, but it does not claim completeness nor does it deny a Canadian bias. Attempts were made to discuss all topics equally thoroughly, but it is likely that for some readers the book is too detailed in some respects and too superficial or incomplete in others. Papers published after spring 1974 are not included in the text.

As the English system of measurement is still being widely used in English-speaking countries and in aviation around the world, it is used in this book. A table to convert measurements from the English to the metric system is given at the end of the book (page 190).

Snow geese at their staging grounds

Chapter one: Masters of the air

BIRDS AND BIRD MIGRATION

This chapter presents information on birds in general and on bird migration in particular. It is written to provide background material for those readers who are unfamiliar with birds. It is by no means a short text on ornithology but rather a collection of bird data pertinent to the bird strike problem. Birds have attracted man's interest throughout history, and publications about them are innumerable. A great deal of the following information was obtained from textbooks that are listed in the reference section (325, 402, 429).

BIRDS IN GENERAL

BIRD CLASSIFICATION AND IDENTIFICATION

For many years scientists have been subdividing the birds of the world into a system that shows their relationship to one another. This system of orders, families, genera, and species is organized in such a way that it reflects the evolutionary history of birds, and thus expresses current views of what kind of birds (or species) came first and what came later. Thus, penguins, loons, and pelicans, for instance, are considered "old" in the sense of having developed early in biological history. Most of the "new" species are placed in the order of the "Perching Birds," which comprises all "songbirds" or passerines, e.g., swallows, crows, pipits and finches. All told there are about 8,600 recognized species of birds, some of which are further subdivided into subspecies or races. The Canada Goose, for example, has several races. Adult birds of the largest race, the Giant Canada Goose, range in weight from 7.8 to 16.4 pounds (184), whereas the smallest race, the Cackling Goose, does not weigh more than 4 pounds (157).

Birds are recognized by their size, shape, and sounds, and are identified mainly by their distinctive markings, called "field marks," and by comparison with similar species. Field guides, containing illustrations and descriptions of the different species, usually indicate the key field marks. Guides for different geographic areas can generally be purchased from local bookstores. Although field guides are helpful, it takes much experience to identify with certainty many species in field conditions.

BIRD NUMBERS

Nobody knows and few have ventured even to guess how many birds there are in the world. Fisher (149) estimated the world population at 100,000,000,000 birds, "that is, there are probably a hundred thousand million rather than a million million or ten thousand million." This famous ornithologist put the bird population of Britain at about 120 million. It has been estimated that each fall about 5,000 million passerines migrate from the west and central Palaearctic (i.e., Europe, and Asia north of the Himalayas) to wintering areas south of the Sahara in Africa (298).

Two "educated guesses" of the land bird population in the United States arrived at 5½ and 6 thousand million birds at the start of the breeding season (150). For the state of Illinois the total summer population in 1957 was estimated at 49 million birds, with 54 million during the preceding winter (168). A recent census for North Dakota arrived at a state-wide breeding population of 26 million pairs (381).

Common Starlings are common indeed. At their winter roosts, they gather in large numbers. Nine roosts in England varied from 15,000 to 1,400,000 birds (388). In 1890, small numbers of starlings were brought from Europe and liberated in New York and since then their numbers in North America have increased tremendously, with their range gradually spreading over most of the United States and southern Canada (165). A starling roost in Pennsylvania in northeastern US was estimated in the late 1950's to contain at least one million birds (144).

Total numbers of blackbirds and starlings wintering in the US are currently estimated to be as high as 500 million. During winter blackbird roosts in the southeastern US often contain several million birds each (358).

The Herring Gull population of New England has increased from just over 10,000 in 1900 to about 70,000 in 1955, with population numbers stabilizing in the last two decades (134). In certain areas in Europe gulls have increased in numbers to the extent that they are now considered a pest, polluting reservoirs of drinking water, befouling buildings, causing havoc in bird sanctuaries set up to protect other species, and colliding with aircraft (72). The total wintering population in England and Wales was estimated at some 1,500,000 gulls (198).

Waterfowl have probably been studied more intensively and their numbers estimated more often than any other migrating species, because of hunting and the resulting need for proper management practices. Some 90,000 swans, more than 3 million geese, and 50 million ducks migrate in spring into Canada (98). Apart from these millions of geese and ducks, some 300,000 Sandhill Cranes (98) and a few thousand White Pelicans visit Canada in the summer. Waterfowl numbers in Europe and Asia are considerably smaller than in North America (21,299).

Leaving aside the reliability of the above figures, it is clear that large numbers of birds occur at or over many localities during at least part of the year.

BIRD SIZES AND WEIGHTS

Birds show an enormous variety in size and appearance, ranging from small hummingbirds with a length of 2½ inches to the Wandering Albatross with a record wingspan of 11.5 feet. There is a similar spread in weight: one hummingbird species weighs less than a tenth of an ounce, whereas a male Trumpeter Swan has been recorded at 38 pounds (324).

In some game bird and passerine species, the male is heavier than the female, but females of some predator species (e.g., falcons) consistently outweigh the males (324). In species with a wide distribution, populations in colder regions tend to have a larger body-weight (and thus a higher volume-to-surface ratio) than those in warmer regions. This is an adaptation of the birds that helps them maintain their body temperature. The Screech Owl, for instance, weighs 6.3 ounces at the northern limit of its range, but only 4.4 ounces at the southern (439).

Birds vary in weight during the course of the day, weight being greatest late in the afternoon and least late in the night. Apart from daily fluctuations, migratory birds also show well-marked seasonal fluctuations, storing fat before migration. "In certain migrants that may make a non-stop journey of 1,000 miles, as much as 100 per cent of the lean body-weight of the bird may be added as fat" (324). Thus a one-ounce bird may put on as much as one ounce of migratory fat.

Several scientists have published records of bird weights. These records, however, generally pertain to a certain area or certain bird groups only (e.g., 307). At the request of the Canadian Associate Committee on Bird Hazards to Aircraft, a list of bird weights based on published and unpublished records was prepared for various bird groups (waterfowl, waders, gulls, birds of prey, some passerines, and a few other groups) (238). There are plans in the United Kingdom to prepare an improved and more extensive list. Some bird weights are given in Appendix 1-1.

THE SENSES OF BIRDS

Some idea of the stimuli that birds can perceive through their sense organs is needed for understanding bird behaviour and how it may be manipulated, for example in attempting to drive birds from an airfield.

Vision

Birds have a highly developed optical system: they have relatively big eyes and great visual acuity. Nocturnal birds in particular have eyes which permit them to operate at low light intensities. Birds have good colour vision, although some species have less sensitivity for the blue end of the spectrum (394). There is no indication that birds can perceive light of either shorter or longer wavelengths than man can (in other words, they too are unable to see ultraviolet or infrared light).

Hearing and balance

These senses are discussed together because the sensory apparatus for both is in the ear. Hearing in birds approaches that in man in the precision of the auditory functions that have been measured, such as acuity of hearing, pitch discrimination, and ability to locate a source of sound (327). Most attempts to demonstrate in birds a sensitivity to ultrasonic sounds (higher than about 20,000 Hertz or cycles per second, the upper limit in human hearing) have been unsuccessful.

Birds maintain their balance in the same way that other vertebrates do: the ear has a structure to detect angular and rectilinear movements of the head. When a flying bird receives information of such movements, it immediately uses the muscles of its wings and tail to counteract any roll, pitch, yaw or stall. A hovering kestrel demonstrates how well this system works.

Smell

Can birds smell? After at least 100 years of experiment and debate, this question has not been settled. To date, data remain fragmentary and contradictory, but it can be assumed that some birds have a sense of smell which is poorly developed. Its influence on the behaviour of birds is little understood (429).

Taste

Birds have chemo-receptors called "taste buds" at the base and side of the tongue and in the softer regions of the palate. Compared to mammals, birds have very few taste buds, and one would therefore expect this sense in birds to be inferior to that in man. The taste sense of birds has been tested in food preference and other experiments, and the evidence strongly suggests that pigeons and domestic fowl have a well-developed sense of taste and are able to discriminate at least some substances that taste salt, sour, and bitter (137). Nevertheless, chickens select their food with a preference for shape and colour, with taste having little or no effect. Studies on taste in birds are important in the search for repellents that might be sprayed on crops and so reduce bird damage.

Touch

The sense of touch encompasses the capabilities to feel pressure, change of temperature, and pain. Birds have no special sense organ, but nerve endings have been found in the bill and oral cavity, and in the skin at the base of some feathers (429). Little is known about this sense in birds.

Other senses

Birds are able to navigate, (i.e., they are able to orient themselves in the absence of landmarks previously known to them). After centuries of observations and decades of experiments, it is still incompletely understood how birds manage to navigate. In 1964 Matthews concluded: ". . . it now appears that two types of navigation are shown by birds; a simple distance and bearing type and a more complex form of grid navigation. Both appear to be based on celestial clues, using either the sun position or star pattern. These conclusions appear certain in the case of the simple compass orientation and plausible but 'not proven' so far as true navigation is concerned" (285).

Since then, radar studies have shown that birds are able to maintain a straight course when migrating in or between layers of clouds thick enough to impede greatly a view of the sky or the ground (173). Also, new experimental results have indicated that the earth's magnetic field may be important in orientation (437, 438). It is still unknown how birds sense magnetic forces.

Despite the results of these and many other sophisticated studies, our understanding of birds' navigating abilities is far from complete. In recent years there has been a trend to emphasize that birds may have multiple forms of orientation behaviour. Nevertheless, it was pointed out that for many birds, such as homing pigeons that show good homeward orientation within a few minutes after release under overcast skies in unfamiliar territory, or migrants flying straight and level through opaque clouds, there is not one adequate explanation (173).

BIRD MIGRATION

The seasonal comings and goings of birds must have amazed primitive man as they still intrigue modern man. Several books have dealt with bird migration, some of which are mentioned in the reference section (132, 172, 353).

THE ANNUAL CYCLE

In the northern hemisphere many bird species are migratory, that is, they migrate in spring from southerly wintering areas to northerly breeding areas and return in autumn to the wintering grounds. In a country at moderately high latitudes, Denmark, for example, one discerns: (1) summer residents that breed in Denmark, but winter farther south, (2) spring and fall transients breeding to the north but wintering to the south, (3) winter visitors that winter in Denmark but breed farther north, and (4) permanent residents present the year round. To complicate matters, if all populations of a wide-ranging species move several hundred miles south in the fall, different populations of the same species may be found in any one place at different times of the year. Furthermore, now and then accidentals, casuals, and rare birds are noticed.

In general, when migrants arrive at their breeding grounds, they make their nests, lay and brood their eggs, rear their young, molt their feathers, and fly back to the wintering grounds. This is, of course, the well-known annual cycle of main bird activities, although the exact sequence of events varies from species to species. Because of the seasonal variations in bird distributions a particular problem that occurs in January may have solved itself in April, whereas by July a completely different problem may develop.

During certain periods of the year, bird migration and local flights of densely flocking species pose a serious threat to flight safety. These topics will therefore be discussed in more detail below.

METHODS OF STUDYING MIGRATION

The enormous technological development of the last few decades has made it possible to study migration in much greater detail and on a much larger scale than ever before. Where a biologist once had only a pair of binoculars and a notebook, he may now use a specially designed radar in combination with multi-track tape recorders.

Bird banding is useful to obtain an idea of the general distribution of birds, especially during migration and at the wintering grounds. It is also possible to get information on the rate of progress during migration by capturing birds the next day or a few days after they have been banded (256). The banding technique was given a new dimension with the introduction of coloured neck bands with numbers (in contrasting colours) large enough to be read with a spotting telescope from a distance. This type of band, which can only be used on big birds (e.g., swans and geese), provides the biologist with a means of observing individuals in the field.

Visual observations of migrating birds can be made with the naked eye, binoculars, or a telescope. By lying on their backs and looking up through binoculars, Dutch biologists discovered the "secret" migration of Chaffinches at some 1,650 to 2,640 feet above ground level (409).

Birds observed in the beams of a lighthouse provided some information on nighttime migration (133), but it is likely that under certain weather conditions such light beams significantly alter the birds' normal behaviour. A more universally applicable method is based on counts of birds silhouetted against the full moon, as observed with a 20-power telescope (277). Comfortably installed, an observer can watch the moon continuously for about an hour before his eye becomes tired. Moonwatch data can be used to calculate the directions of migration and the number of birds crossing a mile of "front" per hour. The "front" is a horizontal line at right angles to the main direction of migration.

The "ceilometer" technique allows visual observations of low-level migration during any night that it does not rain and thus has wider applicability than the moonwatch technique (159). Briefly, this method employs a strong beam of light which is aimed vertically into the sky and birds flying through the beam are counted by an observer with a telescope, lying some five to ten feet away from the light source.

Visual observations need not necessarily be made from the ground. Most aircraft observations of birds in flight (289, 293-295) are made during the daytime, but Bellrose counted birds at night as they flitted through the beams of auxiliary landing lights installed on the landing struts of his Piper 180 (33). As this type of survey was inviting the very hazard that this book is about, he did not fly when he expected heavy waterfowl migration (34).

A completely different way to study nighttime migration is to identify and count the flight calls of migrants that pass unseen in the dark. This technique has been improved and automated through the use of a microphone, parabolic reflector, and tape recorder (167).

Rapid progress in the field of bio-telemetry has made it possible to attach a small battery-operated transmitter to the back of a bird that can thus be tracked by car or airplane equipped with a movable antenna (110, 360).

In the last 15 years or so radar has yielded a wealth of new information on many aspects of migration. It has given us more information about heights, speeds and directions during migration, and has provided a clearer picture of how birds react to certain weather conditions. In 1967 Eastwood produced his timely book *Radar Ornithology*, giving an explanation of radar bird detection and a review of published results up to 1966 (138). A bibliography on the detection of birds, bats, and insects by radar, published in 1964 (302) was updated in 1969 (303).

Radar may well become an important tool in bird strike reduction programs; consequently radar bird detection for use in air traffic control is discussed in Chapter 6.

TIME OF YEAR AND TIME OF DAY

Academic debate on exactly what causes migration still goes on, but it is generally agreed that length of daylight has much to do with it. The consensus is that the seasonal change in the day length both in spring and autumn produces physiological changes in birds, which thus become ready to migrate. If the bird is ready it will fly as soon as the weather is favourable (132). Since seasonal changes in daylight are the same from year to year, birds of the same species are likely to migrate at approximately the same date each year. This is true especially in spring.

German ornithologists introduced the terms "instinct-birds" and "weather-birds" to indicate that some species, such as the Cliff Swallow and Baltimore Oriole, arrive at their destination about the same date every year, regardless of the weather, whereas the arrival of others, such as the Woodcock, Snipe, and Common Starling, is determined by the weather. In practice, however, it is often impossible to decide to which group a given bird belongs (132). Thomson (401) did not mention these categories when he summarized: "For any species or population of a species the dates of migration are, as a rule, remarkably constant. It is only within narrow limits, apart from weather movements of hardy migrants, that the dates vary in accord with the meteorological characteristics of particular years."

Some species migrate by day, others by night, and still others by both day and night. In the last category belong the loons, geese, ducks, gulls, terns, and shorebirds. Most small songbirds fly only at night (unless they are making a long over-water flight), probably because they feed during daylight and are safer from predators at night. Hawks and eagles fly only during the day, and then only when good updraft conditions prevail. Swifts and swallows migrate by day, probably feeding on the wing. Both moonwatch and radar data have shown that nighttime migration starts shortly after sunset, peaks around midnight, and stops before dawn.

DIRECTIONS AND ROUTES

The notion that birds migrate north in spring and south in fall is a sweeping generalization. In fact, there are only a few species that fly along a north-south axis, even though for most species their wintering grounds are at lower latitudes than are the breeding areas. All species have their own breeding and winter range, and their own routes from one to the other. In some species the birds follow different routes in spring and fall.

Many birds change direction en route. Spring migration routes of the eastern population of the North American Whistling

Swan are shown in Fig. 1-1 as an example of such changes in direction.

Radar has shown that at night, migration occurs over a broad front, but that some daytime migrants tend to become concentrated along coastlines or valleys under certain weather conditions.

Moonwatch and, particularly, radar observations have shown that many bird species make long-distance over-water flights. Even small birds can cross the Mediterranean (297) or the Gulf of Mexico (277). There is evidence that some shorebird species cross a good part of the Atlantic Ocean when migrating from southeastern Canada to the West Indies and South America (334,436).

For soaring birds the crossing of a large stretch of open water is a serious obstacle, as is clearly shown in the migration routes of the White Stork (Fig. 1-2). These storks fly around rather than across the Mediterranean because they need rising air currents, which are prevalent over land but absent above the sea (353). The Crane is similar in general appearance to the White Stork, but even though it is a good soarer, it is also capable of sustained flapping flight. Cranes migrate by day and by night and are known to cross the Kattegat Channel between Sweden and Denmark (5), and the Mediterranean (353).

HEIGHTS OF MIGRATION

Over a period of 15 years that involved 12,000 hours of flying and about 1,750,000 air miles, Mitchell recorded 217 sightings of birds flying between 500 and 12,000 feet above sea level (asl). Most sightings occurred under 4,000 feet asl. His observations showed great variations even for a single species. Ten records for the Lapwing varied from 600 to 12,000 feet and 32 observations of Swifts ranged from 600 to 6,000 feet asl (293-295).

A compilation of height observations from all over the world, made mostly from aircraft, included a spectacular record of a Lammergeier over Mount Everest estimated at 24,000 to 25,000 feet and a Condor in the Andes at about 19,800 feet asl. Geese were seen up to 9,000 feet and Wood Pigeons up to 12,000 feet asl (289).

All these pilot reports pertain to casual daytime observations. Only Bellrose has used an aircraft for a systematic study of the height distribution of birds at night (33). In the spring and fall about 50 per cent of the birds between ground level and 5,000 feet above ground level (agl) occurred at the 500-1,000-foot height

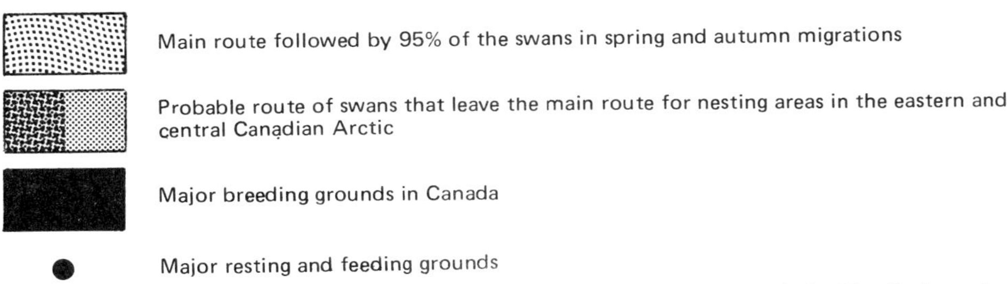

Fig. 1-1: Routes followed by the eastern population of the North American Whistling Swan during their annual migration across Canada and the United States (after 360).

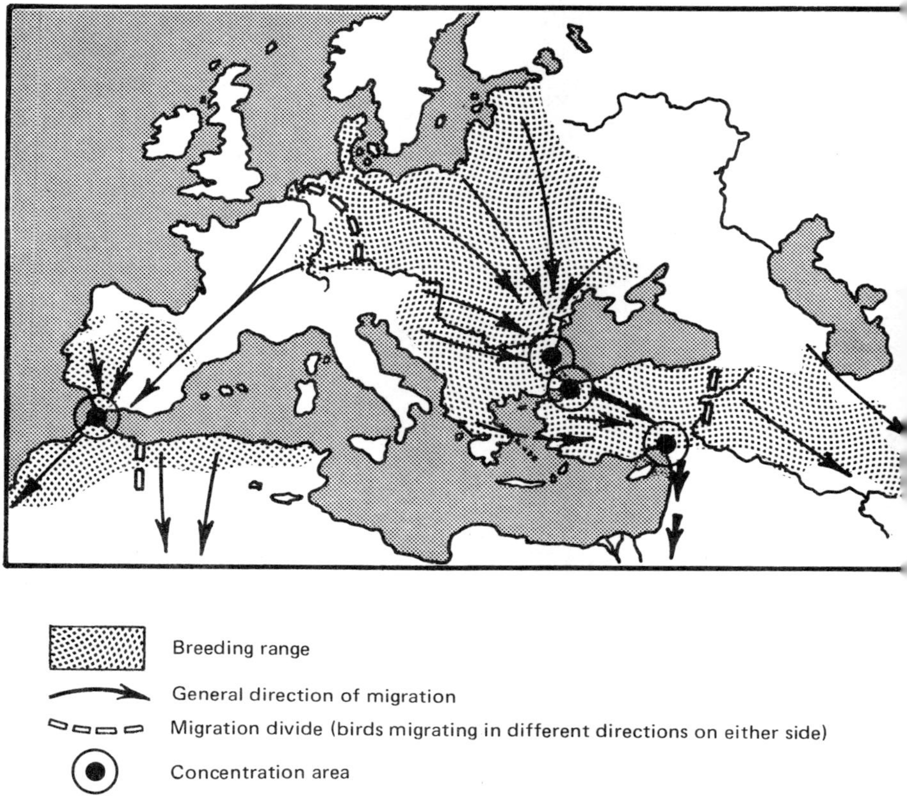

Fig. 1-2: Fall migration routes of the European White Stork (after 353).

band. With the exception of one night, few small migrants were seen at 5,000 feet or above.

A wealth of height data has been obtained in recent years with radar (Appendix 1-2). These studies show that in general about 90 per cent of all migration occurs below 5,000 feet agl, but that some birds, especially flocks of shorebirds on long-distance flights, can be encountered at altitudes of up to 20,000 feet agl.

Radar observations also showed that the height of migration may vary from night to night, and sometimes even from hour to hour. It is not yet completely understood what factors influence the height of migration. The altitude of a particular bird at a particular hour probably depends on the species, the time and locality of departure, the destination, and, especially, the upper-air weather conditions. Birds apparently dislike flying in fog, in rain, and in or between clouds although they have been reported doing so occasionally (173). Unless clouds are too high for birds to surmount, they attempt to fly above an extensive overcast. When unable to climb above the cloud deck, the birds are usually concentrated in the layers immediately below the clouds (35). As birds often migrate with following winds, it seems plausible that they fly at those heights where they have the best tailwind conditions (29, 332, 377). However, optimum flying altitude may also be determined by the aerodynamic and physiological characteristics of the species and thus by factors such as temperature, relative humidity, oxygen pressure, and air density (80, 322).

The high altitudes at which migrating birds have been reported indicate their abilities to fly effectively at low air and oxygen pressures, as well as at low temperatures. The temperatures at the heights at which the birds are flying are of interest to the aircraft engineer who is designing bird-proof windshields. Most materials become more brittle at lower temperatures; windshields of commercial aircraft are therefore heated to maintain the strength of the materials. Nevertheless, bird strikes at low ambient temperatures are likely to cause more damage. Information on the temperatures where birds are flying is scarce. Although the results of a radar study in England (140) suggest that the freezing level acts as a flight ceiling, this is not always the case. In fact, in Canada many flights in late fall start when ground temperatures are below 32° F (0° C). An excellent radar study in Switzerland (80) showed that in 27 of 40 migration periods at least 10 per cent of the birds flew at heights where temperatures were below freezing, the highest birds flying at air temperatures as low as 5° F (−15° C). Some of the highest birds over Primrose Lake, Alberta, were also found at temperatures of 5° F (56). Although the species flying at these low temperatures were not identified, it is likely that flocks of shorebirds and waterfowl were involved, rather than single flying passerines.

FLIGHT SPEEDS

The speed of birds in flight has often been exaggerated in the past, but actual speeds are impressive. Of course, flight speed during migration is lower than that attained in forced flights for short distances. Over the years, many records have been based on observations of birds that were seen from car or airplane, or that were followed with theodolite or rangefinder (289). These records show a wide range in groundspeeds, even for a single species, e.g.,

Common Starling and Carrion Crow (Appendix 1-3).

Airspeeds of birds are hard to determine because one needs to know the groundspeed, direction, and altitude of the birds as well as the speed and direction of the wind at that height. The following airspeeds of migrating birds were obtained from radar studies:

Small birds	US	21 — 30 mph	(135)
Small passerines	North Sea	about 23 mph	(252)
Waders	North Sea	35 — 46 mph	(252)
Lesser Snow Goose	Canada	28 — 41 mph	(376)
Lesser Snow Goose	Canada	31 — 38 mph	(55)
Crane	Sweden	37 — 52 mph	(5)

Airspeeds may be affected by the altitude of the birds (at higher levels, the air is thinner, causing less drag) (322), the "motivation" of the birds (37), and the direction and force of the wind (birds flying with tailwinds may reduce their own efforts as wind speed increases) (29, 80).

RATE OF MIGRATION

In general, birds do not migrate between their summer and winter grounds in one long flight, but in a series of short flights with ample time en route to feed and replenish their fat reserves. Geese are known for their long flights, but once they have arrived at a staging area (or stopover), they may spend a week or more before embarking on the next stage of their migration. Small passerines make shorter flights during their migration, with more stops to feed, but they do not linger for weeks in any particular area.

Many species begin their spring migration with short flights, which become longer as the birds get closer to their destination. A study of all species travelling up the Mississippi Valley in the United States indicated an average rate of advance of 23 miles a day. From southern Minnesota to southern Manitoba, 16 species maintained an average rate of migration of 40 miles a day. From there to Lake Athabaska in northern Canada 12 species travelled an average of about 72 miles a day and 5 others travelled to Great Slave Lake with an average daily progress of 116 miles a day (270).

Autumn flights of North American waterfowl are rather different from the spring migrations. Several species, and particularly the Mallard, remain in the north until they are literally frozen out. The birds gather in huge flocks on the last open water, and after the passage of a strong cold front there is a spectacular exodus and a high, long flight to the south (36).

FLOCKING AND GROUPING

Many bird species are gregarious, especially during migration. Most of the larger migrants fly in flocks. Swans, geese, ducks (especially in the fall), cranes, storks, pelicans, loons and "waders" all migrate in more or less organized flocks. Many of the passerine daytime migrants (e.g., finches, starlings, crows) fly in flocks, whereas most of the passerine nighttime migrants (the great majority of all species) fly either singly or in loose groups. Such "loose groups," ranging between 2 and 12 birds, would probably extend over an area 100 or 200 feet across (313).

In Britain it was found that although some of the night migrants were probably true flocks, others, and perhaps the majority, were "pseudo-groups," a result of the radar characteristics rather than the birds' mode of flying (141). Over Switzerland, 92% of nighttime migration consisted of single birds, about 5% of pairs of birds, and 3% of flocks with 3 or more birds. In contrast, only about 35% of daytime migration consisted of single birds (80).

DENSITY OF MIGRATION

The density or intensity of migration is often expressed as the Migration Traffic Rate (MTR) or the number of birds crossing one mile of front per hour (the front being a horizontal line perpendicular to the main direction of migration). In North America, a major effort to make simultaneous moonwatch observations across the continent was carried out on three consecutive autumn nights (278). Nearly 1,400 volunteer moonwatchers contributed their data to this project. The calculated Migration Traffic Rates varied greatly: on locations where there was little or no migration, the MTR's ranged from 0 to 100; but in areas with heavy migration, peak MTR values of 10,000-20,000 and, in the southern US of 30,000-40,000, were obtained. At Baton Rouge, Louisiana, the highest MTR was 61,500 birds per mile of front per hour (311). Assuming that the birds had a groundspeed of 40 mph, the last figure would mean a density of about 1,500 birds over one square mile. As nighttime migration usually occurs over broad fronts, it is obvious that on a single night of heavy migration millions of birds are on the wing, and the area involved is virtually covered by a blanket of birds.

Radar studies yielded similar results. A peak MTR of over 15,000 was calculated during heavy nighttime migration over east-central Alberta (53), and an average MTR of 3,680 was obtained for the hour around midnight for 10 nights without weather changes in Switzerland (80). An estimated 3 million birds took part in a heavy migration over Hertfordshire, England, on one spring night (187), and each year there are at least 10 nights on which more than 5 million birds cross the radar screen at Cape Cod, Massachusetts (316).

EFFECT OF THE WEATHER

In most regions of the northern hemisphere, migration is not a regular, even flow of migrants throughout the migration season. Bad weather usually suppresses migration almost completely, and if poor weather conditions persist for several days the number of birds physiologically ready to migrate builds up. When the weather finally becomes favourable, a heavy migration usually occurs (196). On the other hand, when favourable conditions persist for several days the number of birds ready to migrate is reduced, and low densities are reported at the end of such a period.

Rain, sleet, snow, fog, and strong headwinds are usually considered as unfavourable weather conditions, whereas tailwinds or calm conditions are deemed favourable. High temperatures in spring and low temperatures in fall are often associated with heavy flights. However, weather factors are highly interrelated and mathematical treatments are required to determine what weather factor or combination of factors may be correlated with heavy or light migration (333).

LOCAL FLIGHTS

At their breeding and wintering grounds, and even at migration stopovers, some bird species make regular local flights. Because they are a known hazard to flight safety, the main types of local flights are summarized schematically:

Spring and early summer

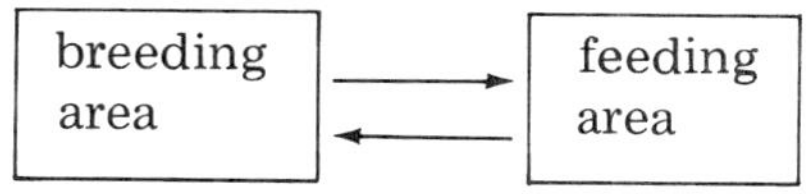

Late summer through winter

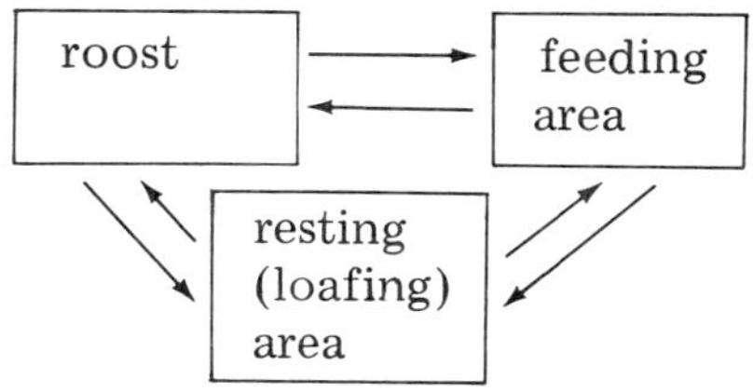

After the breeding season, Common Starlings and Red-winged Blackbirds congregate in enormous flocks that spend the night at favourite roosts. Their morning departures from the roosts are spectacular and the timing of such flights can easily be predicted as they usually start about sunrise (139).

Many big cities and their immediate vicinity provide attractive feeding, loafing, and roosting areas for gulls. Gulls are scavengers and will feed at a garbage dump, near a sewage outlet, or at a fish-processing plant. They loaf in open pastures or on the concrete surfaces of an airport, and they spend the night on a nearby lake or other body of open water. Thus, gulls often have routine (and largely predictable) daily flight patterns during certain periods of the year. These patterns are relatively stable, although they may change with the season or with weather conditions.

Shorebirds that feed on exposed intertidal flats concentrate during flood tide in large numbers on "high-tide roosting sites" where they preen, loaf, and sleep until low tide.

Geese on migration may spend a few weeks in traditional staging areas, where they often pass the nights on lakes, riverbanks, or flooded fields and make regular daytime feeding flights to agricultural fields.

During the breeding season pelicans are known to make long flights to obtain fish for their young (39, 276). White Pelicans breeding on Great Salt Lake, Utah, US, make round trips of up to 100 to 150 miles. As pelicans like to soar, they carefully select the right wind conditions when air freighting the baby-food. They either pick tailwinds or they choose a route that provides them with good updrafts (276).

These are just a few examples of a great variety of local bird movements which have the following features in common: they may occur close to airports; they may involve large numbers of birds, often in flocks; and they are fairly regular and thus largely predictable.

FLOCK DENSITY

Chapter 3 describes the attempts to make aircraft "bird-proof." Turbine engines are highly vulnerable to bird damage. Thus, aircraft designers need to know how many pounds of bird meat and bird bones their engines should be able to withstand. For example, if an airliner runs into a large flock of starlings, what are the chances that any of the engines will ingest 1, 2, 3 or more birds, and how much time is there between successive birds? An engine may be able to withstand a lot of starlings one at a time, but may give out if 10 birds are ingested simultaneously.

Little is known about the three-dimensional spacing of birds in flying flocks. One published report (424) describes a photographic method of estimating densities of bird flocks in flight. Densities, expressed as the number of birds per 1,000 cubic feet (10 x 10 x 10 ft.) of air, were calculated for 60 species ranging in size from swallow to swan. Flock densities, some of which are given in Appendix 1-1, varied from 2 or 3 birds per 1,000 cubic feet for swans, to 1,200 birds per 1,000 cubic feet for small sandpipers.

During a study in the UK, based on observations on bird densities on and around runways, it was calculated that a total of 120 pounds of bird weight might strike an aircraft with a 100-square-feet frontal area during takeoff or landing (386).

Apart from encounters with large flocks of geese or swans in midair, the worst risk for airliners is probably during lift-off and the first stages of climb, when a lot of engine thrust is required to get the fully loaded airplane to its cruising altitude. If an aircraft on takeoff runs into large flocks of starlings, gulls or shorebirds, there is little room for manoeuvring and even a brief malfunction of the engines as a result of bird ingestion may prove disastrous. In Canada, tests are underway with air photography to determine the densities of flocks of gulls and sandpipers a few feet above the ground. These tests are designed to simulate conditions when a flock of birds, feeding or loafing on the runway, takes to the air just ahead of the approaching airplane.

BEHAVIOUR OF BIRDS WITH RESPECT TO APPROACHING AIRCRAFT

BIRD BEHAVIOUR IN GENERAL

The behaviour of a bird includes what it does and how it responds to its surroundings (325). Students of behaviour often differentiate between innate behaviour, which is instinctive, and learned behaviour. Innate behaviour occurs without previous experience or learning. It often presents itself in so-called "fixed-action patterns," and a particular stimulus at a particular time will "release" or trigger this behaviour. These releasing stimuli are provided by "releasers," which consist of certain shapes, colour patterns, sounds or movements. An example of a releaser is the red patch on the lower half of the beak of the Herring Gull, which elicits food-begging behaviour in the chick (408).

As in man, learned behaviour in birds results from practice or experience. The limits of what a bird can learn are determined by its innate behaviour. Although it is convenient to speak of two distinct classes of behaviour, instinctive and learned behaviour are usually so fused that, in general, a particular action cannot be considered the result exclusively of inheritance or of learning. The process of most learning in birds may take place through "imprinting," through trial and error, or through habituation. "Imprinting" is part of the learning process limited to a brief period in early life.

As aircraft are not usually part of the bird's natural environment, birds have to learn, either by themselves or from their parents or flock members, how to behave safely with respect to approaching aircraft. Thus, it is natural that young or otherwise inexperienced birds will be frequently involved in accidents, because they have not yet learned to stay away from dangerous objects, such as powerlines, shotguns, vehicles, or aircraft.

Before examining the behaviour of birds with respect to aircraft, collisions between birds and another device for fast transportation — automobiles — are briefly reviewed.

BIRDS KILLED BY CARS

Each year large numbers of birds are killed by road traffic. Many birds become casualties because they feed from road surfaces.

Heavy rain "washes" many insects out of the air onto the ground, and may bring worms out of the ground onto the road surface. Insects, frogs, and birds killed by traffic attract scavengers such as crows, that in turn may become casualties. Partridges, finches and sparrows, in particular, are killed as they dust-bathe or take grit from roads. Other birds may be struck by vehicles as they drink or bathe in rainwater puddles. Another attraction of roads for birds may be that they offer a place to keep their feet dry (147). In England, low-flying birds — such as finches, sparrows, game birds, thrushes and swallows — are often killed while crossing narrow roads bordered by thick hedgerows (207).

Birds that frequently use roads usually have a keen awareness of road traffic and react well in time. However, when such experienced birds become preoccupied with other matters, they may react too late. In the breeding season, birds are busy establishing and defending a territory, building the nest, and feeding the young. This is precisely the period when many birds are killed by automobiles. Immature birds also are often involved in collisions. A study in England (207) showed that over half the birds killed during July, August and September were young birds.

The author was involved in a car collision with a Great-horned Owl that shattered the windshield. The owl had caught a rabbit and was reluctant to leave the road without its quarry. It slowly came off the ground with the dead rabbit in its talons, and finally dropped the animal just before it was hit by the car. This behaviour may explain why relatively large numbers of owls, hawks, and falcons are found among the road kills in Alberta, Canada (430). The reason for a large number of owls killed by cars may also be that many hunt from dusk to dawn, and may thus become disturbed or blinded by the lights of a car. A report of a Barn Owl shattering the windshield of a motorcycle after dark mentioned that the habit of owls swooping toward a moving light, especially during the breeding season, may account for a number of casualties (207).

In conclusion, it appears that car-bird collision risks are governed by the behaviour, experience, and activity of the birds as well as by the speed and visibility of the cars.

BIRD BEHAVIOUR WITH RESPECT TO AIRCRAFT AND GLIDERS

Observations on the behaviour of birds in the presence of aircraft were published as early as 1916, when "flying men" began to compete with birds for airspace (143). However, apparently little has been published since that year and no systematic study has been undertaken to investigate how birds react to approaching aircraft. Obviously the birds' behaviour would show an enormous variety in response (ranging from panic to indifference), depending on factors such as the size, shape, colour, visibility, noise, height, and speed of the aircraft; the species, sex, age, health, and experience of the birds; the time of day and time of year; and the weather conditions. Some published information on the subject is reviewed here.

Birds sitting on the ground at airfields

In 1943, an RAF Officer reported some of his observations made during seven years of flying. Birds breeding or feeding on airfields paid little attention to the movements of aircraft on runways and taxiways. He was able to approach a brooding Lapwing to less than five yards when he taxied his aircraft. However, if a man walked beside the aircraft the bird would fly off its nest at 100 yards (106).

Observations on the behaviour of gulls in Britain showed that they usually do not settle on or near a runway in use when there is much air traffic. When gulls happened to be close enough to be startled, their reactions often showed an element of panic. Instead of flying away at right angles to the path of the aircraft, they sometimes performed a swerving, erratic flight for up to 50 yards in the path of and away from the airplane (similar behaviour of gulls was noted when they were flushed by an automobile). Other species (Lapwings, Rooks, and Common Starlings) did not show this panic flight in front of aircraft (28).

Other studies in the UK (77) reported that Lapwings are variable in their reactions to aircraft. Birds were often seen sitting right on the edge of the runway and completely ignoring the airplanes, whereas other birds might be disturbed at 200 yards away. Oystercatchers on the grass at the edge of the runway might walk a yard or so from an approaching plane, but they rarely took flight, and when they did, they would fly only a few yards away from the airplane before landing. On one occasion, Oystercatchers were seen on the tarmac while two aircraft took off, and the birds narrowly avoided a strike. Wood Pigeons showed considerable variation in their behaviour. It seemed that local birds feeding along the runway were not disturbed by aircraft, whereas migrant birds were readily alarmed while feeding close to the runway. Starling flocks were found to be more often disturbed by aircraft, the closer they were to the runway. They tended to take flight when the airplane was still distant and fly away from it at a low altitude. Rooks flew up in response to 12 of a total of 31 aircraft movements, and were more apt to do so when close to the runway (78).

Black-headed and Herring Gulls resting on the rocks at the perimeter of Nice Airport in France were completely indifferent to air traffic, but when the gulls were resting directly under the line of traffic, they rapidly left their spot during takeoff by an aircraft so that it would not fly over them (119).

Observations at Vancouver International Airport in Canada showed that Dunlins (Red-backed Sandpipers) behave in a particularly dangerous way. When pursued by a falcon, they bunch up in a dense mass and fly in a zig-zag fashion directly away from the falcon. Dunlins flying across the runway when an aircraft is approaching are likely to react to the airplane in the same fashion as they do to the falcon. That is, they tend to form a tight flock and zig-zag directly down the runway away from the airplane rather than veering off to the side (265).

Birds sitting on the water

Wildlife biologists often make aerial censuses of waterfowl resting or feeding on water. Atlantic Brant Geese flush from water while the aircraft is still more than 1,000 yards away, but Snow and Canada Geese allow an aircraft to approach within a few hundred yards or less. Both diving and dabbling ducks flush at a much shorter distance than do geese (128).

Birds in the air

Although one pilot reported that birds in the air usually wheel leisurely away on approach of an aircraft (106), this does not appear to hold for waterfowl. Canada Geese seemed to have three ways of reacting to small aircraft: (1) attempting to outfly the aircraft, (2) side-slipping out of the way, and (3) folding wings and plummeting earthwards. Wood Ducks, American Widgeons, and Pintails also showed the free-fall dropping behaviour, but Mallards and Black Ducks continued to attempt to outfly the aircraft at distances of perhaps even less than 50 yards (128). The free-fall behaviour was also reported for Greater Snow Geese by an alarmed observer, who witnessed these five- to ten-pound birds "falling as snowballs all around the plane" (300).

It may be that this free-fall dropping behaviour only occurs when the aircraft is below the birds. When birds and aircraft are at the same level the behaviour appears to be different. One pilot, flying a light aircraft against the direction of but at the same height as numerous flocks of Lesser Snow Geese, reported that all flocks veered around his aircraft, but continued migrating in the original direction. In a few cases a flock split in two parts and reformed behind the aircraft (24). Another pilot encountered a flock of Bean Geese that immediately veered away. On another occasion, he pursued a flock of Bean Geese in a JU-52 (with flaps down and at half throttle). These birds were not intimidated and kept their height and direction, until the wing of the aircraft moved into their formation. They then turned away and descended (352). A study in the US reported "... that small birds such as blackbirds, swifts and swallows, make last-second plunges to avoid aircraft, but ducks and geese take evasive flight at much greater distances ahead of the aircraft" (33). The behaviour of waterfowl appears to vary with circumstances: although ducks, and particularly geese, are in general wary of man and his machines, they appear to be less frightened by aircraft when making a migratory flight.

If birds are scared and try to avoid an aircraft, the success of their evasive action may largely depend on whether they are able to manoeuvre and accelerate rapidly. Long-winged birds with a slow wing beat and a preference for gliding (such as gulls) are probably less capable in this respect than short-winged birds with a fast wing beat (such as small shorebirds).

Birds of prey may not avoid aircraft, but even attack them, especially in remote regions where aircraft are a novelty. In 1942, a Golden Eagle attacked a military aircraft; the bird broke its wing, but the aircraft suffered no damage (352). In 1971 a light aircraft crashed near Port Hardy on Canada's west coast after colliding on descent with a Bald Eagle (94). There were no survivors so it is not known whether the eagle attacked the airplane, or was simply reluctant to leave a good thermal. When pilots report that they were attacked by an eagle or hawk, the report may be in error because of an optical illusion. Due to the high closing speed between birds and aircraft, a soaring bird that was not alarmed by an approaching aircraft might seem to the pilot to have attacked.

There have also been reports of eagle attacks on gliders. One pilot approaching an eagle nest saved his machine and his life by a well-aimed pistol shot (232). In another case buzzards were observed to attack and break model gliders at about 2,000 feet altitude. The same birds also approached a motor-powered model, but abruptly turned away and did not attack it.

A glider was also used by Pennycuick to study the soaring performance of some East African birds. He reports that Tawny Eagles showed a positive reaction to his glider, and, in fact, when he joined a thermal beneath one, "the eagle would often dive down on the glider at an alarmingly high closing speed, taking up formation behind the inside wing with a spectacular braking manoeuvre, often well within the wing tip. On several occasions tawny eagles dived at the glider from straight ahead, usually missing it by a narrow margin; one, however, collided head-on with the starboard wing, and was last seen tumbling earthwards, apparently dead. No damage was suffered by the glider on this occasion, but these birds did constitute a significant hazard" (323).

Martial Eagles showed a somewhat similar aggressive behaviour, and Pennycuick suggested that eagles may regard a glider as potential prey. White Pelicans and White Storks on the other hand often showed signs of alarm and tried to dodge when his glider was still at a great distance, suggesting that these birds were being preyed upon in flight (for instance by Martial Eagles) and that they considered the glider as a potential predator. Two vulture species neither attacked nor avoided his glider and often as many as five birds would fly in close formation just behind the tail, with several more trailing along behind (323).

All the pilot reports mentioned above are based on daytime observations. At night, most small birds showed little concern for a small lighted aircraft. Larger birds behaved differently. In clearing the propeller, most small birds passed close to the fuselage; middle-size and large birds were more likely to pass farther out on the wing; and large birds passed near the tips of the wing. During 49 nights of flying in a light aircraft (Piper 180) over east-central US, three small birds and no large birds were struck. Radar observations made simultaneously confirmed that small birds were neither attracted nor repelled by the lights of the aircraft (33). Although radar offers good opportunities to study the behaviour of birds with respect to aircraft, it has not been widely used for this purpose. In Canada, bird flocks have been detected on the screens of the Precision Approach Radar at Winnipeg International Airport. On one occasion a flock of migrating Snow Geese backtracked twice to avoid landing aircraft and then proceeded to cross the approach sector (179). Similarly, in the US flocks of ducks and geese were seen on radar to deviate from their flight course when aircraft passed nearby (33).

To avoid an approaching aircraft, a bird must be able to

anticipate the flightpath of the aircraft. With straight-flying aircraft this is probably not too difficult, but serious problems arise when an aircraft changes height or makes a turn. Research in the USSR has shown that this ability to visualize a flight path is weak in gulls, ducks, pigeons, and predators (234). In France it was also noticed that many strikes occurred on turning aircraft (261).

We may conclude that the behaviour of birds with respect to aircraft has not been thoroughly studied and, consequently, is poorly understood. Incidental observations made by pilots of slow aircraft indicate that the bird's behaviour varies with species. Virtually nothing is known where fast aircraft are concerned, but it is likely that the reactions of the birds would be basically much the same to fast as to slow aircraft.

The skin and windshield of this CF-104 aircraft were torn and broken by a collision with a flock of birds.

Chapter two: Hard facts about soft feathers

BIRD STRIKE STATISTICS

What is a bird strike? For the purpose of this book, any contact between a moving aircraft and a bird (or group of birds) is referred to as a bird strike. For the bird the result of colliding with an aircraft is usually fatal. For the aircraft the results can vary from a light smear of blood, a dent or hole in the structure or a cracked window, to a disabled engine that might lead to a crash.

REPORTING, ANALYZING, AND PUBLISHING OF BIRD STRIKE STATISTICS

Bird strikes were rarely reported in the early years of aviation. When they became more numerous and did more damage, some airline companies and military flight safety groups began to compile statistics to determine the seriousness of the problem, and to get some idea of how and where to approach it.

REPORTING PROCEDURES

The prescribed way in which bird strikes on civil aircraft are to be reported in Canada may serve as an example of how bird strike statistics can be collected (Fig. 2-1). The aircraft captain is responsible for reporting all bird strikes and "bird near-misses." He uses a tick-off reporting form which is in his flight bag. The form is sent to the Aviation Safety Bureau of the Ministry of Transport (MOT), and a copy is held by his airline company. If a strike occurs during takeoff or landing, the pilot informs the control tower (ATC) by radio. This information is relayed to the airport ground maintenance staff, which checks the runway for dead birds. If birds are found, a bird strike report form is filled out, and a bird wing and other feathers are sent to the Canadian Wildlife Service (CWS) for identification.

Fig. 2-1: Bird strikes that occur in civil aviation in Canada are reported according to the plan outlined in this chart. Reporting is initiated from the left side.

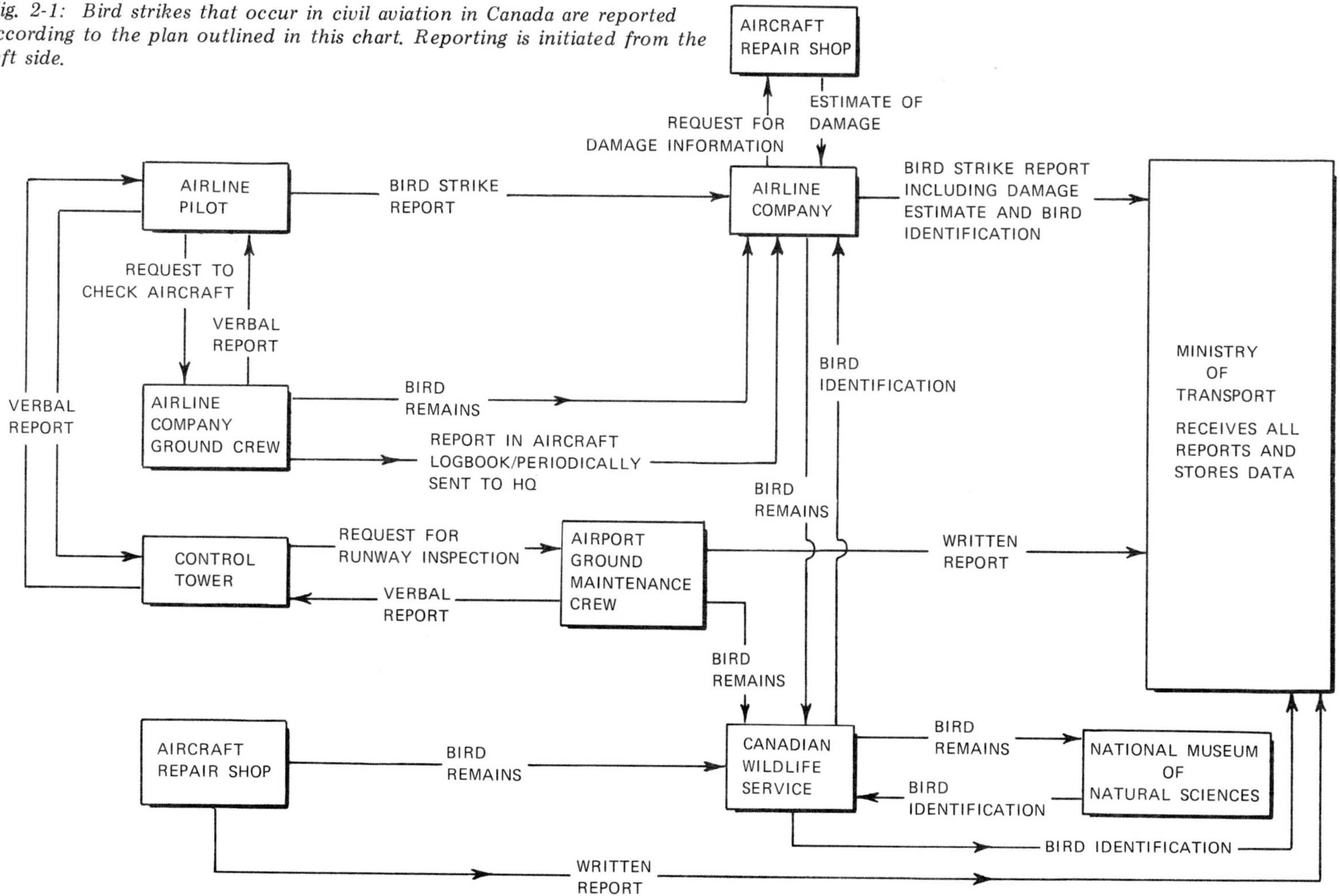

Sometimes the flight crew is unaware that a bird strike has occurred. In such cases the bird collision may be detected after the aircraft has landed by the airline's ground crew, which does a "walk around inspection" on all company aircraft that use the airport. If they detect a bird strike, they can either tell the pilot (if he is still there) or report the incident in the aircraft logbook. Bird remains are sent to the airline company, which forwards them to the CWS.

If the tower controller suspects that a strike may have occurred, because he sees birds flying near an aircraft, he may query the pilot and may also request a runway inspection. Furthermore, runway inspections are carried out routinely each day and all birds found dead are reported to MOT and the bird remains are sent to CWS. The airport manager has the responsibility of ensuring that all known and suspected strikes occurring on his airfield are properly reported.

It may also happen that a strike is not detected until an aircraft undergoes a periodic overhaul in the maintenance and repair shop. In such an event, any bird remains found are sent to CWS, and MOT is notified. The CWS sends the bird feathers to the National Museum in Ottawa, which identifies the bird and reports back.

When a strike has occurred, the airline asks the repair shop for an estimate of the damage and cost, and passes this information to MOT.

In order to encourage the strike reporting process, MOT has sent reporting forms to all owners and operators of turbine-powered aircraft (except helicopters). As reporting is not mandatory, a friendly letter accompanies the forms, urging pilots to use them. Bird strikes on non-turbine-powered aircraft will, in general, be reported only if they are "significant."*

It can be seen from Fig. 2-1 that detection and reporting of strikes has many steps and requires much voluntary cooperation for its success. Unfortunately this cooperation is sometimes lacking. Since 1973, MOT has entered all the reported information into a computerized data storage, the reporting form having been designed with this system in view. Although MOT has not published the data, it forwards its strike summaries annually to the

*According to the ICAO Airworthiness Committee 7th Meeting Working Paper 294, a bird strike is "significant" if there is: catastrophe, loss of life or destruction of aircraft; torn metal skinning of any part of aircraft; deformed primary or secondary structure; cracked or shattered wind-screen; or damage to engines sufficient to cause shut-down or unrecoverable loss of power.

headquarters of the International Civil Aviation Organization (ICAO) in Montreal, where they are compiled with similar reports from other countries. These compilations are then distributed to all member states.

Although bird strike reporting has recently improved as a result of extensive publicity, strike records are still far from complete, because many people do not follow the reporting procedures. Until a few years ago the situation was considerably worse. Sometimes airlines or other aircraft operators had kept extensive bird strike records, but were reluctant to publish the information because it might have an unfavourable effect on air travel. Air Canada and KLM Royal Dutch Airlines were among the first airlines to publish their bird strike records, and many major companies now release their strike figures regularly.

ESTIMATING COSTS

It is usually difficult to assess accurately the costs of repairing or replacing damaged components on an aircraft. Accurate costs are possible only for "non-lifed items" (i.e., parts that do not have a stated lifetime). A jet engine, for example, has an expected life, and if it is destroyed halfway through this period the loss is approximately half the cost of a new engine.

When a strike on a jet engine is suspected, an examination is required (Fig. 2-2). For complete internal checking, the engine has to be taken out of the aircraft and examined in a specially equipped repair shop. Even if no damage can be found and the engine is reinstalled in the aircraft without any repairs, the operator has obviously incurred a cost in time and money, but these costs in terms of "man hours" and "overhead" are often difficult to evaluate.

Commercial passenger flights delayed or cancelled due to a bird strike may involve additional costs to feed and lodge the passengers and possibly to pay for alternate transportation.

Like oceanliners, big airliners lose money when they are idle. If a B-747 has an engine breakdown and has to wait for a replacement engine, the operator may lose tens of thousands of dollars a day. A grounded aircraft brings no revenues, yet normal operational costs continue (interest on the investment of well over $20 million, salaries to crew, and other overhead). In addition, the company may incur other costs such as parking fees for the aircraft.

Military organizations, when estimating costs of an aircraft

Fig. 2-2: The technician is examining a B-747 turbine engine, following a bird strike. A special instrument called a boroscope is used to inspect hidden areas.

rendered unserviceable by a bird, may use only the number of hours of repair required to make the aircraft operational again. The Canadian Armed Forces now use the following four categories of damage:

A Category — The aircraft is destroyed, declared missing, or damaged beyond economical repair.
B Category — The aircraft must be shipped to a contractor or depot-level facility for repair.
C Category — The aircraft must be flown to a contractor or depot-level facility for repairs; or repairs are carried out by a mobile repair party; or a major component has to be replaced.
D Category — Damage to any component can be repaired within field level resources.

In this system of classifying damage the cost factor is not of prime importance, and it is extremely difficult to arrive at damage figures in terms of dollars.

IDENTIFICATION OF BIRD REMAINS

Identification by bird experts of remains found on the runway, on the aircraft, or in its engine, establishes the bird species and sometimes the age of the bird involved. This tells the engineer the probable weight of the bird (Chapter 1), and provides the biologist with the first information on which to base a biological bird control method. For these reasons, proper identification of feathers is essential. Different species usually have different habits, which may be important to those who must plan preventive action. Identification of feathers can best be done by biologists in museums where large skin-collections are available.

Sometimes the remains of birds ingested by jet engines are so charred that no feathers or other material remain for identification by the unaided eye. However, microscopic analysis of samples scraped from the engine surface has almost always shown the presence of tissues that are characteristic of birds. Even if the microscope yields no information, it is possible to analyze the scraped material for its amino acid content, which would confirm that a collision with a bird (or perhaps a bat) had occurred. This technique, pioneered in Canada by LaHam (253, 254), has been of great help to investigators of aircraft accidents. As the J79 jet engine of CF-104 aircraft was often involved in bird strikes, the

Canadian military authorities issued an illustrated "J79 Bird Strike Manual," showing bird damage and bird deposits on various engine components, and describing LaHam's technique (101).

A more recent method offers even greater possibilities. Using techniques that have been developed in clinical chemistry, it is possible to identify the bird species involved in a strike from a minute amount of blood smear found on an aircraft. Certain proteins (albumins) in the blood are specifically identifiable with bird species. Albumin can be isolated from bird blood samples, purified, and stored, then albumins from an unknown bird can be compared with them and identified by immuno-chemical techniques, not described here. Just as a "feather file" (collection of bird skins) is needed to identify a feather, a collection of albumins from different bird species is needed to identify the albumins in a smear of blood (154).

RELIABILITY AND USEFULNESS OF BIRD STRIKE STATISTICS

The foregoing description of how strikes are detected and reported, and how their damage is estimated, will serve to emphasize that the accuracy and reliability of strike statistics may differ for different countries, airline companies, and national air forces. This makes it difficult to arrive at a reasonable assessment of the bird strike problem.

Tabulations of injuries, crashes, and deaths caused by birds are essential in evaluating the seriousness of the problem, but reported numbers may be less than the actual numbers. Data on yearly strike numbers, or the number of strikes per 10,000 take-offs and landings (strike rates), may indicate trends, but they seldom allow statistically meaningful conclusions, because the sample size (i.e., the number of strikes) is usually small.

In summary, it is difficult to obtain records of all bird strikes, and it is virtually impossible to get all details for those strikes that are reported. This does not mean that the reporting of strikes is a useless exercise and a waste of time and money. It means that caution is needed when conclusions are drawn from bird strike statistics. Even incomplete data can provide a rough idea of the frequency of bird strikes, of their costs (in terms of human life and dollars), of the relative vulnerability of aircraft components, of the bird species involved, and of the effectiveness of certain anti-bird measures at a particular airport. Obviously, the better the statistics and the more thorough the analysis, the more solid the conclusions. To obtain the best possible picture of the bird strike problem, all bird strikes should be reported as fully, accurately, and promptly as possible.

TYPES OF DAMAGE RESULTING FROM BIRD STRIKES

DIRECT DAMAGE

Most strikes occur on the forward-facing parts of an aircraft (Fig. 2-3). A bird strike on the aircraft skin may result in a smear of blood, or in dented, pushed-in, crumpled, or torn metal. In a strike on a sloping surface there is usually only a glancing blow, but when the collision occurs more nearly at right angles to the surface, then bird debris and torn metal may result.

When a bird hits the blunt part of the leading edge of an aircraft wing, damage such as is shown in Fig. 2-4 may occur. An example of heavy bird impact on the tail structure is shown in Fig. 2-5.

Bird impact on the nose cone has resulted in dented fuselage (Fig. 2-6) and actual penetration (Fig. 2-7). In the latter example a Common Loon crashed through the radome of a Buffalo aircraft

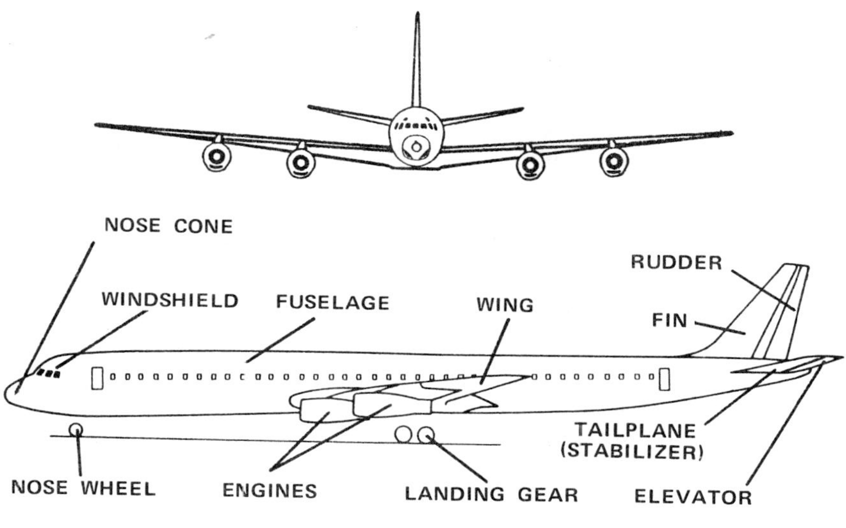

Fig. 2-3: Some components of a jet airliner.

Fig. 2-4: Damage caused by the impact of an unidentified bird on the wing of a Tutor aircraft.

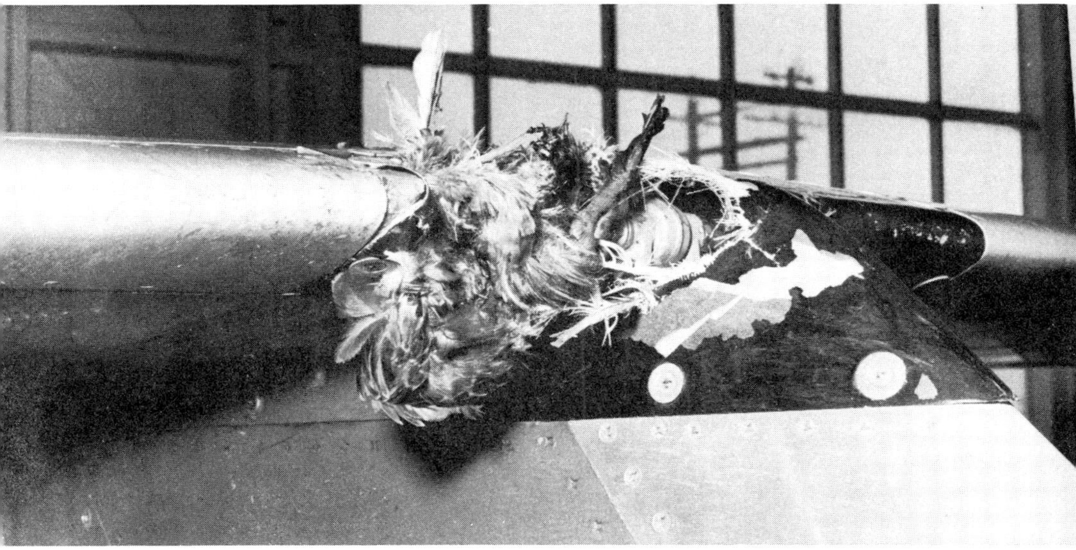

Fig. 2-5: Damage done to the vertical and horizontal stabilizers when a Pintail duck struck the tail of a Tutor aircraft.

Fig. 2-6: A Herring Gull struck the nose of the Varsity training aircraft and pushed it in.

and bent the radar antenna. The radome is a protective cover for a radar antenna and is made of a special material that allows radio waves to pass through. On the Buffalo the radome also functions as part of the nose cone.

Strikes on a windshield may smear it, or crack, shatter, or completely destroy a panel. In the worst cases pilots have been exposed to flying pieces of glass, bird debris, and a strong draft. In at least one case a pilot has been killed. Two examples of windshield damage are shown in Fig. 2-8.

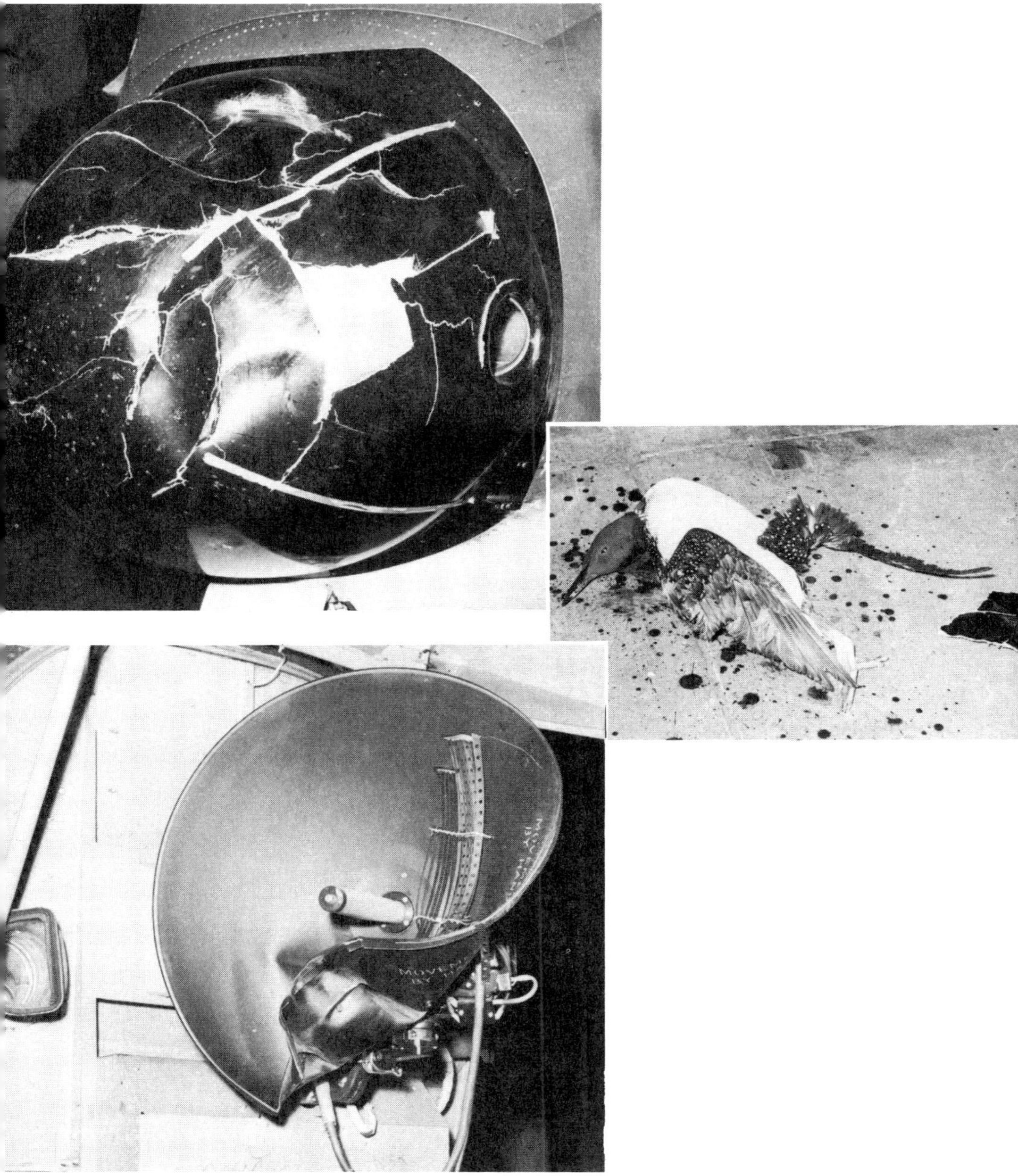

Fig. 2-7: When a Common Loon collided with a Buffalo (CF-218) aircraft, the nose cone was broken and the radar antenna inside the cone was damaged.

Fig. 2-8: Bird impacts shattered the windshields of these CF-104 aircraft.

The propellers on piston-engined aircraft are strong and, although often struck, they are rarely damaged by birds; further, the rotating propeller blades tend to protect the strong piston engines by chopping up the birds, and strikes causing severe damage to piston engines have not been reported. Turbine engines, however, are a different matter altogether. They are usually exposed and are quite vulnerable to bird impact. The rotor and stator blades of the engine's compressor may bend or break on heavy impact (Fig. 2-9) and, in the worst cases are completely destroyed (Fig. 2-10). The design and vulnerability of turbine engines is dealt with in Chapter 3.

Fig. 2-9: (top) This Harrier aircraft was forced to make a crash landing after bird ingestion by the turbine engine. (bottom) The rotor blades were bent and torn by the collision.

Fig. 2-10: Ingestion of a pigeon caused complete destruction of the compressor blades of a turbine engine.

The components mentioned above are the more commonly struck areas, but strikes have also occurred on the underside of the fuselage, the flaps, landing lights, landing gear, etc.

Other things being equal, aircraft with large frontal areas are more likely to suffer bird strikes than those with relatively small forward-facing areas. The so-called wide-bodied aircraft, such as the B-747, DC-10 and L-1011, are of special concern, because they have large frontal areas, and enormous air intakes. At the 1973 meeting of the Bird Strike Committee Europe, a number of serious bird strikes on wide-bodied aircraft were reported, and the following resolution was adopted: "The Committee, having before it information on a series of recent bird strikes, particularly in relation to aircraft equipped with large fan type engines, when these aircraft were deprived of the use of one, two or even three engines, expresses its grave concern and apprehension in regard to the potential danger which exists, particularly around certain aerodromes, and urges States and international organizations concerned to take rapid and effective measures to overcome this serious threat to aviation safety" (40).

DELAYED-EFFECT DAMAGE

As well as strikes that cause direct damage to the component struck, there have been several reported cases of "delayed-effect" damage. Such strikes usually occur on turbine engines.

One example of this problem was a DC-8 that struck a bird on one engine during takeoff from Rome. A visual check showed no apparent damage and the engine was continued in service. About 50 hours later this engine failed during a cruise over Lethbridge, Alberta. Inspection in the repair shop showed blade failure and damage to the compressor. Although there was no proof, it was considered highly likely that the engine failure resulted from bird ingestion (44).

Another accident of this nature shows that care and caution must be continually exercised by the captain of an airliner. A DC-8 with destination Tokyo took off from Vancouver International Airport and struck a large flock of sparrow-sized sandpipers as it lifted off the runway. The cockpit instruments did not indicate any engine malfunctioning or other trouble. The captain reported the strike to the tower, and the ground staff found several hundred dead birds on the runway. This information was relayed to the captain, who then spoke to the head engineer of his company. They decided that the aircraft should return for inspection. After dumping about 20 tons of fuel, the aircraft returned to the airport two hours after takeoff. Examination showed eight bent fan blades in one engine, seven in another, and minor damage to the other two engines. The two engines with serious damage were replaced. The engineers who examined the bent blades thought that two or three hours of operation at cruise power would probably have resulted in blade fracture with disastrous engine damage about halfway across the Pacific Ocean (369).

There have been many aircraft accidents for which the cause has never been established with certainty. It is possible that some of these mishaps were cases of delayed-effect damage following a bird strike.

MULTIPLE CAUSATION ACCIDENTS

A serious strike will do damage to an aircraft, but in addition it may startle the pilot, impair his judgment, or distract his attention. Also, if some part of an aircraft is already malfunctioning, a bird strike may make it more difficult to remedy the situation. An accident which is caused by the combined or successive effects of several factors is usually referred to as a "multiple causation accident." An example of such an accident in which a bird strike was involved will illustrate the point.

During takeoff, a B-747 aircraft collided with a flock of gulls. The crew heard noises that sounded like a bird strike. At the same time they felt a vibration and the instruments showed vibration of No. 4 engine. It was shut down and the vibration ceased. The aircraft returned to the airport, where the runway was wet and slippery. At touchdown, brakes were applied and engine reverse was attempted on Nos. 1, 2, and 3 engines. No. 3 engine malfunctioned and could not be reversed. With two engines reversed on one side of the aircraft but no reverse effect on the other side, the aircraft began to yaw. This made it impossible to continue use of Nos. 1 and 2 engines to any extent. The aircraft skidded past the end of the runway and stopped when the nose landing gear collapsed after striking a slab of concrete. All occupants were evacuated via the emergency chutes, and a few passengers received minor injuries. Damage to the aircraft was substantial.

Visual examination showed no evidence of a bird strike on engines Nos. 1, 2, and 4, but No. 3 had been hit by one or more gulls, which damaged the engine fan blades. Testing of the aircraft systems showed that a component of the thrust-reversing mechanism of No. 3 engine was locked, and that the vibration monitor of No. 4 engine malfunctioned due to a broken wire.

Several factors contributed to this accident. If the crew had shut down No. 3 engine instead of No. 4 (which had not been struck and operated normally), the landing should have been completed without complications. It was logical to assume that the aircraft vibration was due to the apparent vibration of No. 4 engine as a result of a bird ingestion. The wet runway was a further factor (420).

BIRD STRIKE STATISTICS

NUMBERS

Civil aviation

A number of member states of the International Civil Aviation Organization now report occurrence of bird strikes annually and ICAO publishes a summary of the data. Annual numbers of bird strikes for the period 1968-73 are given in Appendix 2-1, which shows that each year hundreds of strikes are reported. Because of variations in reporting procedures, direct comparisons of the data for different countries have limited meaning. However, it is clear that the countries involved in fairly heavy air traffic report a considerable number of bird strikes. For the USSR, the figures reported of between 100 and 200 probably represent only a part of the incidents, and the annual total of actual strikes has been estimated as near 1,500 (229). In many other countries the reporting rate may be somewhat similar.

Especially for countries reporting a fair number of strikes, it is noteworthy that the variation in annual strike numbers is not large, taking into consideration other variables such as fleet size, hours flown, types of aircraft operated, and strike-reporting procedures.

Military aviation

Some figures for strikes in military aviation can be found in Appendix 2-2. These figures show that strikes are numerous and fairly constant, in spite of changes in types of aircraft flown, etc.

The numbers of bird strikes shown in both appendixes appear to be considerable, but a large part of these involve relatively minor damage.

AIRCRAFT CRASHES AND LOSS OF LIFE

Aircraft crashes — some with loss of human life — have followed collisions with birds since the early days of aviation. The first bird strike fatality was recorded in North America in 1912, when Cal Rodgers, the first man to fly across the United States, lost his life after a gull became jammed in the controls of his aircraft, causing it to crash. Since then, about 130 deaths have been attributed to crashes of 9 civil and 65 military aircraft following bird strikes. The data for civil aviation are given in Table 2-1 and for military aviation in Table 2-2. The list of military crashes is incomplete because of security requirements and other reasons.

Table 2-1: Fatal and non-fatal crashes in civil aviation caused by bird strikes.

LOCATION AND DATE	AIRCRAFT	DEATHS	BIRD SPECIES	(REF.)
California, US, 1912	light aircraft	1	gull	(369)
Aberdare Mountains Kenya, about 1955	light aircraft	1	vulture	(255)
Serengeti Nat. Park, Tanganyika, 10/1/1959	light aircraft	1	Griffon Vulture	(403)
Boston, US, 4/10/60	Lockheed Electra	62	Common Starlings	(406)
Maryland, US, 23/11/62	Viscount	17	Whistling Swans	(406)
California, US, March 1963	Beechcraft N. 35	1	Common Loon	(255)
Belfast, N. Ireland, Feb. 1964	Turbulent	1	large seagull	(255)
British Columbia, 2/7/71	Cessna 180H	3	Bald Eagle	(94)
Georgia, US, 27/2/73	Learjet	8	Brown-headed Cowbirds	(419)
Lahore, W. Pakistan,* 15/7/62	DC-3	1	vulture	(406)
Lake Erie, USA, 28/7/68	Fan Jet Falcon	0	gulls	(406)
Khor, Ambadu, 23/7/69	DC-3	0	cranes	(406)
Kazan, USSR, date not given	IL-12	0	duck	(229)

*Not a crash. Co-pilot was killed, but pilot landed the aircraft.

Most of the crashes in civil aviation have followed collision with big birds such as vultures and swans, but flocks of starlings and cowbirds have also been involved. The species that have caused crashes in military aviation have varied in size from the Sandhill Crane and Snow Goose to the little White-throated Swift.

In military aviation, severe engine damage due to bird ingestion has resulted in the loss of a number of low-flying single-engine F-104 aircraft. Fortunately, in most cases the pilot ejected and parachuted safely to the ground.

Table 2-2: Some fatal and non-fatal crashes in military aviation caused by bird strikes.

ORGANIZATION	PERIOD	NUMBER OF CRASHES	NUMBER OF DEATHS	(REF.)
CAF	1964-71	10	0	(100,103)
GAF	1966-70	3	—	(200,203)
RAF	16.11.42	1	2	(407)
RAF, Air Command East Asia	Jan '44—April '45	7	12	(407)
RAF	1953	1	2(or 3)	(255)
RAF	1964-69	3	0	(194)
RNethAF	1956-65	5	0	(50)
RNorAF	up to 1972	1	1	(268)
RSwedAF	1967-72	3	2	(3)
SpanAF	—	2	2	(225)
USAF	up to 1974	19	11	(122)
USNavy	1969-71	4	2	(16)
USSR	1946	1	1	(229)

—Unknown or not given.

There have been crashes, both in civil and military aviation, in which it was suspected that birds were the cause, but such accidents have not been listed in Tables 2-1 and 2-2. One of these accidents involved the crash of a Viscount into the Irish Sea in 1968. No proof of a strike was established, because much of the aircraft was not recovered for examination. However, Bewick's Swans were migrating in the same area that day, and circumstances were sufficiently similar to those of the 1962 Viscount crash (Table 2-1) to point to swans as a possible cause of loss of the aircraft.

NEAR-CRASHES

Occasionally the damage from a bird strike is so extensive that a crash would seem to be unavoidable. Several of these near-crashes have been reported (406). For example, in September 1962 a BEA Vanguard taking off at Edinburgh's Turnhouse Airport ran into a flock of gulls. No. 4 engine failed almost immediately and the other three were also affected. At 3,800 feet No. 2 engine failed and the propeller was feathered. A third engine became critical, but the pilot was able to make a successful landing (406).

Another near-crash was the collision of a CP Air twin-engine B-737 aircraft, coming in for a landing at Winnipeg, and a flock of migrating Snow Geese. Of the several strikes on the aircraft, one caused a heavy dent at the top of an engine air intake cowling and another dented the bottom of the cowling on the opposite side. If the aircraft had been in a slight roll at the instant of impact, both geese would have entered the engines, and the pilot and 82 passengers would have been faced with a crash-landing at night and away from the airport (146).

COSTS

Apart from the risk to human life, the high cost of repairs to aircraft and engines makes the need to reduce the bird hazard urgent. Several countries have published estimates of bird strike costs. The damage sustained from bird strikes by the Royal Netherlands Air Force in the years 1956-65 was roughly estimated at $250,000 a year (50). Norway estimated military aircraft damage at $50,000 to $200,000 a year (268). The ten CF-104 Canadian aircraft lost (Table 2-2) would cost over $13 million to replace. The USAF, operating about 100 bases, sustains an average loss of $10,900,000 per year, that is, the 350 to 400 bird strikes reported each year cost about $32,000 each, or $1.60 per flying hour (122).

In commercial aviation, Air Canada reported that for 2,400 recorded incidents in the period 1959-73, many of them minor strikes, the cost was $2,457,000 (detailed data in Appendix 2-3). BOAC's costs totalled £1,000,000 in the six-year period 1958-63 (12).

It is impossible to set a dollar value on a human life, but clearly there are costs associated with any aviation fatality. Such costs will vary greatly. One estimate of the cost of training a Canadian CF-101 pilot is $315,000 (18). Total costs resulting from the loss of lives and aircraft listed in Tables 2-1 and 2-2 would probably exceed $100,000,000.

This sum does not include the considerable further losses due to injuries, such as loss of an eye and consequent inability to qualify as aircrew, nor the expenditures by airlines and air forces on lost operating time, or repairs to damaged aircraft that did not crash. Such costs can be high, especially if large airliners are involved. For example, a Wood Pigeon damaged the blades of a B-747 engine at Orly Airport, Paris. Because the damaged engine could not be dealt with locally, it took ten days before it was repaired and reinstalled. Reblading, rebalancing, transportation, and installation costs were $130,000. In addition there was a daily cost of some $16,000, increasing the costs to a total of about $300,000 (369).

DISTRIBUTION BY AIRCRAFT PART STRUCK

Information from a few reports of bird strikes on various parts of the aircraft is given in Table 2-3. As strikes on engines are likely to cause the most damage, they are probably reported more frequently than other strikes. This may account for the apparent high incidence of strikes on engines.

The world-wide data for 1970 are based on reports submitted to ICAO by many of the member countries. These data show

Table 2-3: Percentage distribution of bird strikes by aircraft part struck.

	CIVIL					MILITARY
Country	Canada	Canada	USSR	UK	World	US
Period	1961	1961	1969	1961-71	1970	1973
Total no. of bird strikes	131[a]	29[b]	—	741	3,028	327
Reference	(42)	(42)	(234)	(406)	(82)	(414)
Wings (incl. flaps, slots, etc.)	10.7	6.1	25.0	22	2	27.8
Tail (empennage)	3.8	0	2.9	2	1	1.6
Nose (radome)	16.1	12.1	0	28	3	11.9
Fuselage	17.6	9.0	4.1	8	15	0
Windshield	13.0	21.2	13.3	14	13	5.5
Propeller	8.5	0	6.6	0	0	0
Engine (incl. nose bullet & cowling)	24.6	51.5	45.7	21	60	32.8
Landing gear	0	0	1.6	5	6	2.7
Other	5.7	0	0.8	0	0	17.7

a) Propeller-driven aircraft
b) DC-8
— Unknown or not given

remarkably low percentages for strikes on wings and nose, compared to the percentages for various countries individually. Strikes on wings and nose are usually not serious and several countries may not have bothered to report such minor accidents to ICAO.

In general, the percentage of strikes on the various components of civil and military aircraft are fairly similar, except for the wings and windshield.

GEOGRAPHICAL DISTRIBUTION

Collisions will occur wherever birds and aircraft use the same airspace, and bird strikes have been reported from all over the world. Comparisons of strike data for different countries or different airports have little value because of varying procedures in reporting, recording, and publishing strike statistics. However, comparisons between different regions within an individual country may indicate those areas in which strikes are most likely to occur.

Bird strikes have been plotted on distribution maps for Sweden (3) and Holland (50). To evaluate properly the distribution of strike risks, the bird strike rates (number of bird strikes per 10,000 aircraft movements) for different areas should be known, rather than the actual numbers of bird strikes. If certain areas show large numbers of strikes, it is worthwhile to investigate the risks in those areas, taking into account factors such as number and heights of flights over the area, time of day and time of year, etc. For example, many strikes on RNethAF aircraft were reported from the Wadden Sea area, which is famous for its rich bird life. As there are also two target ranges in this area, the high incidence of bird strikes was to be expected. However, it was impossible to relocate the weapon ranges in the densely populated Netherlands (50).

The plotting of bird strike *rates* for different geographical areas will indicate what regions are particularly hazardous, but in general there is not enough information to calculate these strike rates. Thus, effort and money are better put into determining and indicating those areas where birds occur in large numbers, and in recommending that these areas be avoided by aircraft. This approach has been widely used (described in Chapter 6).

DISTRIBUTION BY TIME OF DAY

Most bird species are active primarily during the day. However, many birds go about their business during dusk and dawn, and even through the night. Owls and nighthawks usually do not hunt during the daytime. Many shorebird species feed and fly about at night. Waterfowl, such as Grey Lag Geese and American Widgeons, are reported to feed at night (263, 190).

Apart from these local nighttime activities, there is large-scale migration at night. Many species migrate during the hours of darkness, and on nights with heavy migration millions of birds are on the wing (described in Chapter 1).

Thus, one would expect bird strikes to occur at any hour of the day and night, both at airfields and away from airfields. The facts confirm that this is indeed the case. The distribution of 491 bird strikes in UK civil aviation during 1969-71 (406) was as follows: dawn 17%, day 69% dusk 4%, night 10%. Bird strikes in the USAF during 1972 (310) were as follows: dawn 1.2%, day 65.5%, dusk 4.4%, night 28.9%. Results for civil aviation in the USSR (234) and in Canada (11) also show that strikes occur at all hours of the day.

The above figures pertain to bird strike numbers as such, and not to bird strike rates. If reliable strike rates for daytime and nighttime operations were found to be significantly different, then this might be due to differences in numbers of birds on the wing, bird species in the region, or behaviour of birds with respect to approaching aircraft. The question of behaviour would be interesting, particularly if birds were hit at night more frequently than during the day, other things being equal. Such a situation would indicate a need to give the birds flying at night a better chance of detecting an approaching aircraft so that they could better avoid it. An obvious warning would be strong lights aboard aircraft (discussed in Chapter 4).

Air Canada reported that of all 1971 strikes that took place between 300 and 15,500 feet (the highest altitude reported that year), at least 67% occurred at night. Yet only 20% of Air Canada operations took place at night (46). This information indicates that a disproportionate number of collisions happen at night.

Analysis of time of occurrence of bird strikes may shed some light on certain problems. Such an analysis at Vancouver Airport, for instance, showed that 85% of all strikes between January 1963 and May 1974 occurred during hours of darkness. As up to 10,000 ducks were known to fly from a nearby marsh to this airport at night to feed on the short green grass of the infield areas, it was recommended that the airport have night patrols to disperse the birds (190).

DISTRIBUTION BY TIME OF YEAR

During seasons when more birds use the same airspace as aircraft, more strikes are to be expected. Generally, bird strike data appear to bear this out. From 1956 through 1965 a total of 480 bird strikes was reported on RNethAF aircraft. The monthly strike ratios showed three peaks: one in March (spring migration), one in October (fall migration), and one in July (probably due to the presence of young birds) (50). Analysis of the 94 strikes that occurred in 1973 showed the same pattern (343). An increase in strike numbers during the spring and fall migration was also reported for the CAF (103), the USAF (310), the air force of the USSR (229), and to a lesser extent for the GAF (200). In the UK the bird hazard is much more of a year-round problem: RAF strike numbers for 1964-68 were higher in January than during spring migration and peak numbers were reported for July and August (194).

In civil aviation the situation is somewhat similar. On the North American continent, seasonal peaks in bird strikes were reported for civil aviation in both Canada (11) and the US (239). In the USSR there is a sharp increase from July to November corresponding to an increase in the bird population, followed by the fall migration (234).

In the UK most strikes occurred from August through October, with a second peak for January (the same as for UK military aviation). The monthly bird strike rates for a total of 523 incidents during 1969-71 were as follows (406):

	Incidents per 10,000 movements
January	5.5
February	2.6
March	2.9
April	0.8
May	0.9
June	2.0
July	2.1
August	4.6
September	3.7
October	4.2
November	4.1
December	3.5

The high strike rates for August-November may be due to the presence of large numbers of young inexperienced birds and the still higher rate for January can perhaps be attributed to the numerous gulls that winter in the UK. Gulls frequent many airfields and are involved in many accidents.

The KLM Royal Dutch Airlines reported a total of 294 bird strikes at Schiphol Airport during 1963-67 (247). There was a small increase in strike numbers during March, but most strikes occurred from July through November. During these months strike rates were about three times higher than during the rest of the year, suggesting as possible causes: (1) more birds on the field during July-November, (2) a larger proportion of strike-prone species on the field, (3) a larger proportion of inexperienced young birds on the field, or (4) the presence of strong attractants on the field that made the birds more reluctant than usual to leave the area. During the fall harvesting of sugarbeets grown on the field, Schiphol is extremely attractive to birds, especially gulls (210).

DISTRIBUTION BY HEIGHT

In both civil and military aviation, the height of many strikes is unknown. Statistics for collisions at known heights reported by both civil and military organizations show that the majority of birds are struck at fairly low altitudes. The percentage distribution of these strikes is given below. It can be seen that only a small percentage has occurred at heights above 3,000 feet.

	Height of strikes (above ground level)	(Ref.)
CIVIL		
KLM Royal Dutch Airlines 1963-67	43% at ground level 87% up to 1,000 ft. 98% up to 5,500 ft.	(247)
UK civil aviation 1966-71	86% up to 200 ft. 97% up to 2,500 ft.	(406)
USSR civil aviation 1966-71	42% up to 30 ft. 90% up to 1,900 ft.	(229)

	Height of strikes (above ground level)	(Ref.)
MILITARY		
CAF 1969-71	79% up to 1,000 ft. 97% up to 3,000 ft.	(103)
GAF 1966-68	61% up to 600 ft. 98% up to 2,500 ft.	(200)
RNethAF 1956-65	92% up to 1,500 ft.	(50)
RNorAF 1961-72	71% up to 500 ft. 93% up to 1,500 ft.	(268)
USAF 1970-72	86% up to 2,000 ft. 93% up to 3,000 ft.	(310)

The maximum altitude for a bird strike in the USSR (23,620 ft. above sea level) was recorded on 1 March 1970, over Kazbek, a mountain peak in the Central Caucasus, where an unidentified bird struck the propeller of an IL-18 aircraft. At this time of the year many passerines, storks, and geese are known to migrate over the Caucasus (231). The highest bird strike reported to date occurred at 37,000 ft. asl on an aircraft of Israel's El Al (228).

DISTRIBUTION BY FLIGHT PHASE

During much of the year, most birds fly about at low altitudes — perhaps not higher than a few hundred feet. Thus collisions are most likely to occur at low level, and this is borne out by the strike statistics by flight phase, shown below. In civil aviation the great majority of collisions occur during takeoff (about 38%) and landing (about 41%); consequently, research and development work to reduce bird hazards to civil aircraft should start with problems at and around airfields. However, when strikes occur during the climb-cruise-descent part of a flight they may be more hazardous, because they are more likely to involve large soaring birds or migrating flocks of waterfowl. The percentage distribution of strikes for various phases of flight was:

	Taxi	Take-off	Climb	Cruise	De-scent	Ap-proach	Landing	(Ref.)
Canada 1959-63	0	39	(————	19	————)		42	(43)
France 1967-72	7	40	4	4	5	14	26	(41)
Holland 1963-67	1	35	2	0	2	12	47	(247)
UK 1966-71	1	38	5	1	1	13	42	(406)
USSR	0	37	(————	16	————)		47	(234)

In military aviation, a sizable proportion of flights is carried out at low level. Flight phases can be categorized as (a) at or over the airbase, or (b) away from the airbase. Countries use different criteria to define these two categories, but it is obvious that in military flying many strikes occur en route, that is, away from the airbase, as seen in the following percentage distribution figures.

Air force	Period	At or near airbase	Away from airbase	(Ref.)
CAF	1969-71	62[a]	38	(103)
GAF	1967-68	33	67[b]	(200)
RNethAF	1956-65	32	68[c]	(50)
RNorAF	1961-72	54[d]	46	(268)
RSwedAF	1967-72	40[d]	60	(3)
USAF	1967-72	51[d]	49	(310)

[a]within 5 miles of airbase

[b]above 100 ft.

[c]outside airbase boundary

[d]takeoff, landing and taxi

DISTRIBUTION BY AIRCRAFT SPEED

In civil aviation most bird strikes occur at relatively low altitudes, with aircraft flying well below their cruising speed. In the UK, for example, over 80 per cent of all strikes occurred at speeds between 80 and 160 knots, as shown in Fig. 2-11. A similar distribution of bird strikes was also reported for civil aviation in the USSR, where about two-thirds of all strikes occurred between 83 and 166 knots (229).

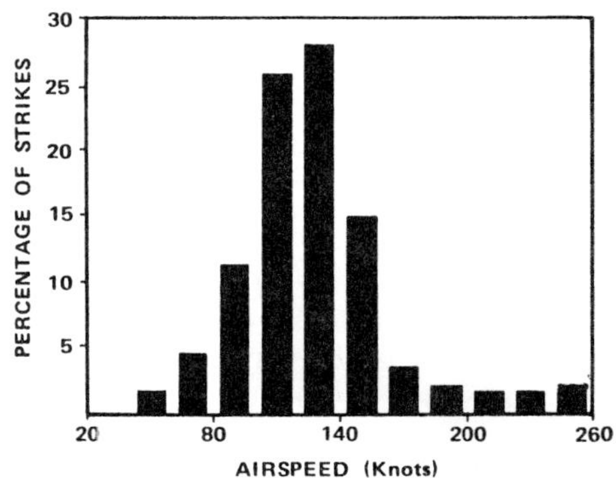

Fig. 2-11: Distribution of bird strikes in relation to aircraft speed (adapted from 406).

BIRD SPECIES INVOLVED

The behaviour of birds with respect to approaching aircraft varies with the species (discussed in Chapter 1). Birds of prey are reported to have attacked gliders and aircraft, whereas waterfowl usually try to avoid airplanes. Birds that usually spend their lives among the trees, such as woodpeckers, are not likely to be involved in a strike unless they are migrating. But birds that frequent airfields are likely to contribute heavily to bird strike statistics. Birds of many species have been struck, and the list continues to grow. In Canada, for example, further bird species are added each year to the list, which totalled 60 in 1973 (see Appendix 2-4) (371). This list shows that birds as big as a Sandhill Crane and as small as a Ruby-crowned Kinglet have been struck by aircraft.

It is, however, more instructive to know what birds are most frequently involved in bird strikes. This information is tabulated in Appendix 2-5, but the data given there are not strictly comparable because of differences in obtaining and identifying bird remains, as well as in analyzing and reporting the results. In addition, the data are a generalization because data for all areas and times of the year are pooled. Nevertheless, they indicate which species are mainly responsible for the bird strike problem in the various countries.

The appendix shows that there is a relatively high percentage of big birds involved in strikes. This may well be due to the fact that bigger birds usually cause more damage, thus increasing the probability that the strike is reported. Also, bigger birds are more likely to leave some easily identifiable remains (feathers, claws, bills) than small birds.

Gull strikes are predominant in North America and Europe, particularly in the UK, where gulls are numerous. For various reasons, gulls are often found on or near airfields. They are good gliders, but they cannot accelerate quickly to get out of the way of approaching aircraft. Furthermore, they sometimes try to outfly the airplane rather than to dodge to the side. As many species are big enough to cause substantial damage, it is not surprising that gulls are the number one pest bird in many countries. At Schiphol Airport, Holland, gulls were found to be involved in many accidents; they were about six times more likely to be killed by an aircraft or its downwash than were lapwings and partridges (383).

Big, slow-flying birds, engaged in activities such as hunting over an airfield, are likely candidates for fatal encounters with aircraft. Birds-of-prey, such as buzzards, typically fall into this category and their relatively large contribution to the bird strike statistics should come as no surprise (Appendix 2-5, raptors).

The aggressive behaviour of some eagle species towards aircraft was mentioned in Chapter 1, and an eagle has been involved in a fatal bird strike (Table 2-1). Although vultures are not known to attack aircraft, they are much too slow to dodge them and have been involved in many serious strikes, particularly in the Far East. Hooded Vultures and Black Kites are a nuisance at Dakar International Airport, Senegal, and at Yuncum Bathurst Airport, Gambia (269). Kites are no problem in Europe and North America, but at Townsville Airport, Australia, the Fork-tailed Kite is the main pest species (262); at Hong Kong Airport the Black-eared Kite has caused problems (340).

Herons are another group of large, slow-paced birds. They frequent airfields to feed on mice and voles. Grey Herons have become a problem at Schiphol Airport, Holland (211), at Kloten Airport, Switzerland (81), and in the USSR (229). A military pilot in Belgium lost one eye when a heron crashed through the cockpit window. The Great Blue Heron caused the crash of a CF-104 at

Cold Lake, Alberta, and disconcerting numbers of this large bird can be seen at Vancouver International Airport, especially during the winter months (182). The related White-faced Heron was a high-risk bird at Auckland International Airport, New Zealand (345).

Cranes are also a known hazard to flight safety. In North America the co-pilot of the USAF T-37 aircraft was killed when a Sandhill Crane smashed through the windshield. In the USSR, the Demoiselle Crane has been involved in strikes (229); in Europe the Crane has had encounters with aircraft (242); and the related Brolga has been struck at Townsville Airport, Queensland, Australia (262). The migration of the Crane across West Germany is carefully monitored (described in Chapter 6).

In Australia, the heaviest flying bird that is a hazard to aircraft is the Australian Bustard. The related Little Bustard has been a problem at Torrejon Airbase in Spain, as well as at Madrid Airport (339).

Both the wild European Wood Pigeon and the tame bird of the pigeon fancier have taken their toll of aircraft engines. As many tame pigeons carry metal leg bands, their ingestion into a jet engine may make a bad strike worse.

Waterfowl (swans, geese, and ducks) are usually wary of man and his flying machines, but their flocks have nevertheless caused extensive damage to aviation. Strikes on waterfowl are more common in the USSR and Canada than in western Europe, probably because of higher numbers of waterfowl compared to other species.

Although such a rare bird as the Osprey has been struck by a Danish jet aircraft (Fig. 2-12), most strikes at airports in Denmark involve common species such as gulls, starlings, and lapwings (236).

Bird problems at a few airports have been studied in considerable detail, and the results of these investigations show, as expected, that some species are a nuisance only in certain areas and often only at a certain time of the year.

Vancouver Airport on Canada's west coast has a host of problem species, one of the worst of which is the small Dunlin (or Red-backed Sandpiper), a winter visitor that occasionally occurs on the airfield in flocks of several thousand birds. They have caused costly (and potentially disastrous) damage to large airliners. As Dunlins cannot be scared away by conventional dispersal methods, experiments are being carried out with novel techniques (see Chapter 5).

Fig. 2-12: An Osprey or fish hawk was killed when it collided with a Danish military aircraft. Its fish catch was not dislodged from its talons.

The Snowy Owl is another example of a bird species that can do much damage during a relatively short time of the year. This Arctic breeder moves southward for the winter, and it is then often present in small numbers at airfields in southern Canada, hunting along runways and taxiways. Inexperienced and unafraid, these birds are likely to be hit by airplanes and have caused a few costly strikes, particularly at Toronto International Airport.

The USAF has a special problem at Whiteman AFB, Missouri, where an endangered species (the Greater Prairie Chicken) has its traditional "booming ground" (the area in which the birds perform a social display during the mating season) near and on the main runway (286). The USAF has taken measures to cope with the problem without killing the birds (described in Chapter 5).

Of nearly 200 dead birds collected from the runways at Schiphol Airport, Holland, 48% were gulls (mainly Common and Black-headed Gulls), 22% were Grey Partridges, and 12% were Lapwings. Pigeons and starlings were virtually absent from the list (383).

At Airbase Lahr in the southern part of West Germany, resident pheasants were common and were occasionally caught in the drag chutes of jet fighters. There was also a sizable starling roost in a small woodlot alongside the main runway (51).

These examples illustrate the point that the bird strike problem is seldom identical in different areas, and consequently, each problem has to be investigated to determine its peculiar features.

Incidentally, bird strikes have contributed to our knowledge of the height at which certain species may be found during migration. Table 2-4 summarizes some published reports on bird strikes where both the height of the collision and the species involved were known.

Another interesting side issue of bird strikes is the recovery of bird bands through aircraft collisions, and a few of these are tabulated on page 64.

There is little information on the age of birds involved in strikes, but what little is known seems to indicate that young birds are prone to come in contact with aircraft. For instance, at Kingisepp Airport, USSR, six of seven gulls struck between early July and early August were immature birds (234).

Most passerines migrate at night either singly or in loose flocks (as mentioned in Chapter 1). Most shorebirds and waterfowl migrate in flocks, and many nuisance birds at airports, such as gulls, pigeons, and starlings, often move about in groups. In the period 1966-71, of the bird strikes on British civil aircraft 44 per

Table 2-4: Some published records of bird strikes where both the height and the species involved were known.

Height (ft.)	Height (m)	Bird species	Aircraft type	Location	(Ref.)
11,480*	3,500	Mistle Thrush	AN-24	E. of Black Sea, USSR	(231)
5,900*	1,800	Little Owl	LI-2	near Tula, USSR	(231)
5,300*	1,600	Bean Goose	TU-104	Leningrad, USSR	(231)
13,000*	3,960	Shoveler	Britannia	between Bombay and Bangkok	(193)
6,400*	1,950	Evening Grosbeak	Piper Apache	Colorado, US	(306)
10,000*	3,050	Golden-crowned Sparrow	Beech Bonanza	California, US	(291)
8,900**	2,700	Mallard	—	near Frankfurt, West Germany	(431)
21,000**	6,400	Mallard	L-188	Nevada, US	(284)
4,500**	1,400	Curlew	—	Northamptonshire, UK	(120)

* above ground level
**above sea level
— unknown or not given

Bird species	Area where banded	Area where hit by aircraft	(Ref.)
Semipalmated Plover	Magdalen Is., Quebec, Canada	St. John's, Newfoundland, Canada	(84)
Semipalmated Plover	Magdalen Is., Quebec, Canada	St. Maria Airport, Azores	(84)
Evening Grosbeak	Denver, Colorado, US	Boulder, Colorado	(306)
Silver Gull	Geelong, Victoria, Australia	Adelaide Airport, Australia	(423)
Spur-winged Plover	Cambridge, Australia	Hobart Airport, Australia	(423)
"Seagull"	Helsinki, Finland	Oldenburg, West Germany	(200)

cent were caused by flocks and 56 per cent by single birds (406). These data depend entirely on the aircrew judgment of whether the birds struck were individuals or part of a flock. Flocks are usually easier to see and — causing more damage — are likely to be reported more frequently.

It is sometimes difficult to know exactly where a bird strike occurred. In one instance, a Sabena aircraft from Belgium arrived in Montreal, Canada, and on landing reported a bird strike. Later the same week Sabena reported another strike at Montreal. The airline was naturally concerned about the incidents as engine damage was extensive and costly. Remnants of the birds obtained from the engines of both aircraft were identified, and both were found to be the remains of Lapwings, a species not native to Canada but common in Belgium. What had apparently happened was that the birds had been struck on takeoff and had remained against the inlet guide vanes of the engines until a landing was made at Montreal; then, with the altered air flow during engine slowdown and reverse thrust, the birds had become dislodged and had gone into the engines and severely damaged them (250). It might seem surprising that birds would remain in a stable position during a trans-Atlantic crossing, but in a similar experience of a Canadian aircraft a duck remained against the inlet guide vanes for several hours without becoming dislodged (250).

Another unexpected situation occurred after the crash of an airliner in Canada. The crash investigation team found a few feathers in the remains of the aircraft, and it was suspected that birds might have contributed to the crash. Expert examination of the feathers showed, however, that they had been drycleaned and probably came from a pillow (250).

COLLISIONS BETWEEN MAMMALS AND AIRCRAFT

Birds are not the only animals struck by aircraft, although mammals cause few real problems. Collisions with a hare and a fox were reported in Canada (48) and Norway (268). Sheep have been hit by light aircraft on small landing strips in New Zealand (312) and France (261). Elk interfered with operations at an airfield in the USSR; and deer were a hazard at several airfields, particularly at Vandenberg AFB, California, where collisions with mule deer "have occurred with alarming regularity in the past" (114). At another airport in California, a coyote was struck by the nosewheel of an L-1011 (48). There is also an undocumented report of collision with a horse in Scandinavia.

Bats are a different matter, being flying mammals. Bat hazards to aircraft were first reported in 1969 (434, 435). The Mexican free-tailed bat did a great deal of damage to T-38 and T-37 aircraft used at Randolph AFB, near San Antonio in Texas, US. These bats migrate in spring from Mexico into the southwestern United States, where they mainly inhabit large caves. Their numbers build up to an estimated 40 million by August. Each night huge swarms of bats leave their roosts in search of food (airborne insects). Just as do birds, bats show up on certain radars; consequently, a bat avoidance program based on radar monitoring is now in use at Randolph AFB with good success (227). In Australia, the flying fox (a bat species weighing about 1.5 lb.) has been struck when crossing the runways of Townsville Airport, Queensland (262).

CONCLUSIONS

The reporting of bird strikes in full detail is usually a complicated process. Consequently, bird strike records are often incomplete and firm conclusions from bird strike statistics are not always possible.

Each year thousands of strikes occur in civil and military aviation. In worldwide civil aviation there have been at least 12 reported crashes with almost 100 lives lost due to bird strikes. On the military side, some 65 crashes were reported between 1942 and 1973, with at least 35 lives lost. The total financial loss resulting from these crashes is probably in excess of $100,000,000. This figure would be much higher if it included the cost of bird strikes that did not cause crashes.

Bird strikes occur worldwide, usually at low altitudes. Both small and big birds, either singly or in flocks, have been involved in collisions. Most of the strikes with serious damage involved either large birds or flocks of small birds.

Because modern aircraft carrying hundreds of passengers are vulnerable to bird strikes, the bird hazard has become a serious problem in aviation. The following four chapters will review the efforts by engineers and biologists to improve the situation.

This experimental windshield installed on a Tutor cockpit will be exposed to bird impact during tests with a "bird gun." Dummies in flight clothing are placed in the pilots' seats. Results of such a test can be seen in Fig. 3-9.

Chapter three: Stronger structures to bounce the birds

BIRD-PROOFING OF AIRCRAFT AND ENGINES

The bird strike statistics given in Chapter 2 show that aircraft have been struck by birds on all parts and that some of these parts are more vulnerable than others. In the following sections the design and construction of aircraft components are briefly described in relation to the work that has been done to make them more resistant to impact by birds, by what is often called "bird-proofing."

Aircraft structures and components must be sufficiently reliable to meet normal requirements for use and safety. The aircraft must also meet economic requirements; if its weight, for example, is excessive it will be at a disadvantage in competition with other makes.

The parts of an aircraft that are likely to strike a flying bird are obviously the forward-facing areas — the nose of the fuselage, the windshield, the leading edges of wings and tail, the engine cowlings, the engines, and a few less critical parts (see Fig. 2-3). In Fig. 3-1 the frontal silhouettes of a few aircraft types are compared, with solid black indicating windows and engine air intakes. This figure shows the large frontal area of jumbo jet aircraft, and the large proportion of that frontal area taken up by the air intakes.

IMPACT FORCES

Some of the first calculations of the forces involved in bird/aircraft collisions assumed the bird to be "butter soft" and stationary, and the aircraft area struck inflexible and perpendicular to the bird. The impact forces (in pounds) shown on page 71 were arrived at through theoretical calculations (282).

Fig. 3-1: The frontal areas of various aircraft are drawn on the same scale. Windshields and engine air intakes are shown as solid black areas.

Bird mass (lb.)	"Bird diam." (in.)	Aircraft velocity (mph) 50	100	200	300	400	500	600
		Impact force in pounds						
1	3	332	1,330	5,320	12,000	21,000	33,300	48,000
2	4	500	2,000	8,000	18,000	32,000	50,000	72,000
4	5	800	3,200	12,800	28,000	51,200	80,000	115,200
8	6.25	1,280	5,110	20,400	46,000	81,600	128,000	184,000
16	8	2,000	8,000	32,000	72,000	128,000	200,000	288,000

When calculating these forces it was impossible to consider the orientation of the bird with respect to the aircraft at the time of impact. Thus it was necessary to use "effective dimensions," which were chosen somewhat arbitrarily.

Later calculations indicated that when birds are assumed to be semi-rigid rather than butter-soft, impact forces would be 5 times higher. On the other hand, if the aircraft skin could "give" 2 inches, this would absorb about half the impact force. If the angle of impact was 30° rather than 90° the impact force would be reduced by another 50%.

Obviously, the impact forces that are calculated depend on the assumptions regarding the structure and properties of bird and aircraft and their behaviour during impact. Different engineers may arrive at different figures, but all agree that impact forces are impressive.

One quick look at the table will make it clear that to protect the aircraft by sheer structural strengthening would be unrealistic. Bird-proofing will have to involve as much as possible the tapering of aircraft surfaces so that bird strikes will occur at oblique angles, and the provision of "give" where "giving" does no damage. Unfortunately, acute tapering angles are not practical for leading edges of airfoils, the engine cowlings, and windshields (292).

AIRWORTHINESS REQUIREMENTS REGARDING BIRD IMPACT

In the western world, the UK and the US maintain national airworthiness requirements that include bird strike requirements. These are called the British Civil Aviation Requirements (BCAR) and the Federal Aviation Requirements (FAR). Other countries, including Canada, have no bird strike requirements of their own but follow the US or the UK. Aircraft designers throughout the world are eager to meet the FAR or BCAR bird strike requirements, in order to make their product both reasonably safe and competitive in as large a market as possible.

There are two categories of civil aircraft: Transport Category, weighing over 12,000 lb, and "Normal, Utility and Acrobatic Airplanes — limited to nine passengers plus crew or 12,500 pounds in gross weight" (91).

The main airworthiness requirements regarding bird impact are given below.

CIVIL AIRCRAFT — TRANSPORT CATEGORY

Turbine engines

According to British regulations turbine engines must be able to cope with three categories of bird impact. For each category of impact, BCAR have established the conditions under which the requirements must be met.

(1) Large birds, weighing at least 4 lb.

When the front of an engine is struck by a large bird, "no hazardous condition can arise to the rest of the aeroplane as a result of the damage that may be caused."

Conditions: The speed is the maximum True Air Speed (TAS) in normal operational service up to an altitude of 8,000 feet. Note that for this requirement complete loss of power in an engine is not considered to be a hazard.

(2) Medium-sized birds, weighing about 1.5 lb.

When the front of the engine is struck by a number of medium-sized birds "there is no unacceptable immediate or ultimate loss of engine performance, no serious increase of engine operating temperatures or deterioration of engine handling characteristics, over the full range of engine operating conditions, and no dangerous physical damage."

Conditions: Number of birds — one for the first 300 sq. inches of a representative intake area, plus one for each additional 1,000 sq. in. with a maximum of 10 birds. Speed of aircraft — the maximum TAS expected to be achieved during the climb immediately after takeoff. Duration of ingestion — not greater than one second.

(3) Small birds, weighing 2 to 4 oz.

Same requirement as for medium-sized birds, but with the following change in the conditions.

Conditions: Number of birds — one for the first 50 sq. in. of a representative intake area, plus one for each additional 50 sq. in., with a maximum of 16 birds (BCAR).

The US requirements for turbine engines are similar.

Windshields

"On all Transport Category aeroplanes the windows, ahead of and protecting the pilot(s) or other occupants, shall be capable of withstanding the impact of a 4-lb. bird when the aeroplane is flying at the maximum true air speed which is likely to be achieved, in operational service, at altitudes up to 8,000 ft. with the most adverse temperature appropriate to this altitude within the range of climatic conditions for which the aeroplane is certificated." (BCAR).

The US requirements for windshields are similar.

Tail structure or empennage

"The empennage structure must be designed to assure capability of continued, safe flight and landing of the airplane after impact with an 8-pound bird . . ." (FAR).

The UK requirements specify a 4-lb. bird for the tail structure.

CIVIL AIRCRAFT — AIRCRAFT UNDER 12,500 LB.

There are no requirements regarding bird impact for this category of aircraft in the US. UK requirements specify only that the windshield must be capable of withstanding the impact of a bird of at least 2 lb.

MILITARY AIRCRAFT

The airworthiness requirements for military aircraft are mentioned here only briefly. Since 1971, US requirements specify that all transparent areas and supporting structure have a 4-lb bird-proofing capability. For subsonic aircraft the applicable level flight speed is the designed maximum speed in level flight at sea level, whereas for supersonic aircraft the applicable level flight speed will be established for each aircraft at the time of ordering (91).

The requirements for transparent enclosures are contained in Military Specification MIL-A-008865A (USAF) and are under regular review. Other documents deal with the specifications for turbine engines, and they, too, are regularly revised and updated.

At present, it has been proposed that the Canadian Armed Forces adopt the USAF specifications as a "guiding document" until better information, based on analysis of CAF bird strike data, becomes available (91).

BIRD-PROOFING

TURBINE ENGINES

Design

Three types of turbine engines are used in modern aviation: turbopropeller engine, turbojet engine, and turbofan engine (see Fig. 3-2).

The turbopropeller engine consists of a compressor to compress air, a combustion chamber to burn the fuel and expand the compressed air, and a turbine to produce power. The turbine drives the compressor. Excess power is transmitted through reduction gears to the propeller. After passing through the turbine, the exhaust gases issue through a small jet or propelling nozzle, producing a small amount of jet thrust, which is additional to the thrust of the propeller.

The turbojet engine differs from the turboprop in that it is a true jet engine, and derives all its forward thrust from the hot gases being forced rearward. The hot gases leaving the jet or propelling nozzle provide the desired forward thrust.

The turbofan (or turbo bypass) engine is a variation of the turbojet: it has some of the compressor blades lengthened to increase the amount of air drawn through the air intake. This additional air is diverted into an annular casing that surrounds the compressor and is discharged aft of the casing, providing additional forward thrust and a more efficient engine for flying at certain high speeds.

There are two types of compressors: centrifugal and axial flow design. The centrifugal is of solid construction, and seldom damaged by birds. This type, however, is rarely used in modern aircraft.

The axial flow compressor consists of sets of radially

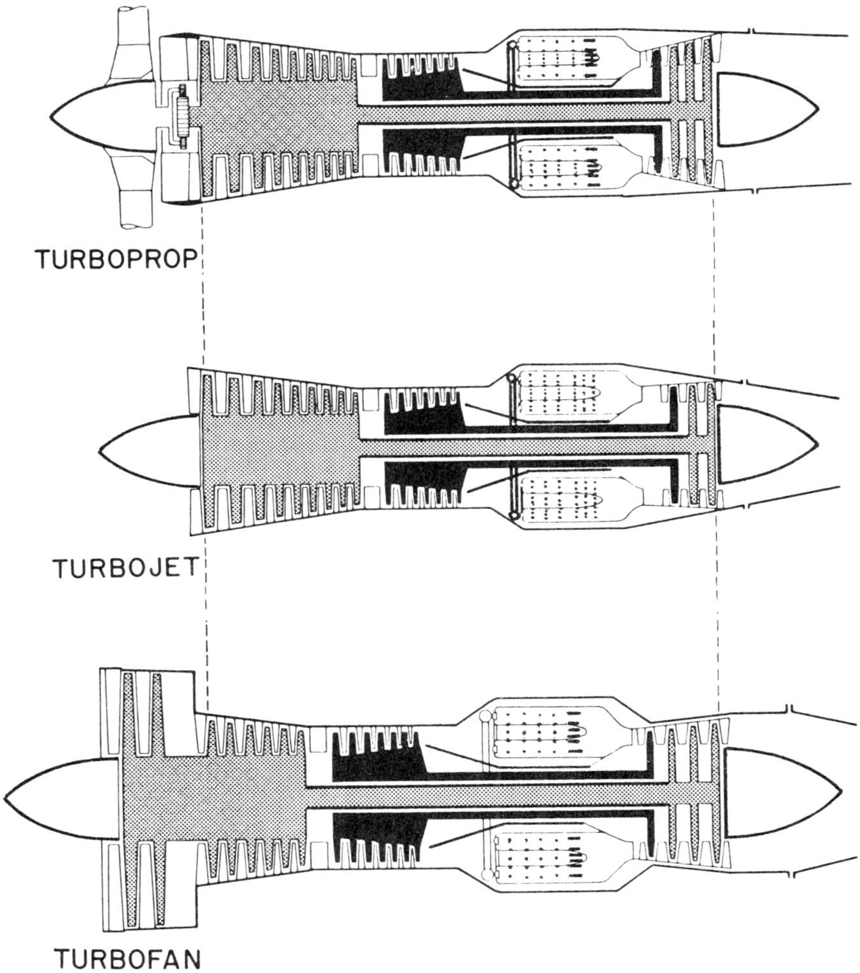

Fig. 3-2: Diagrammatic cross-section of the three types of turbine engines used in aircraft.

arranged blades or vanes, known as stages. Frequently, the first set or stage does not rotate and is known as the inlet guide vanes. Next is the first rotor stage, which consists of a large number of vanes fixed to a hub that rotates. Behind these, fixed and rotating stages alternate, as illustrated in Fig. 3-3. A set of stationary blades is called a stator.

The rotor blades are fastened to the hub (or disc) through the blade root (the so-called "fir tree" root or base, as shown in Fig.

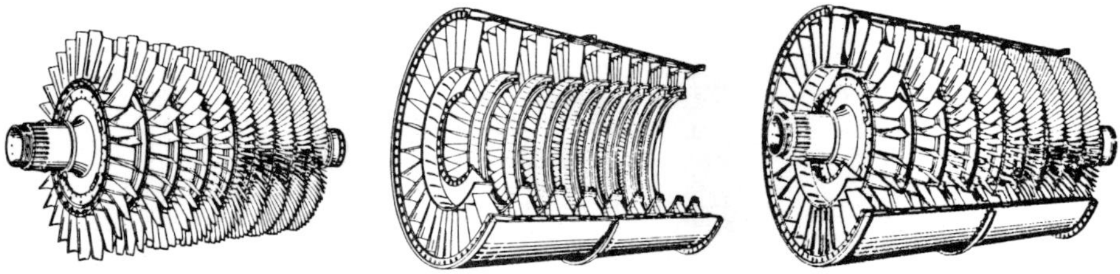

Fig. 3-3: Rotors, stators, and their assembly in an axial-flow compressor.

3-4). The several discs are attached to the compressor shaft by flanges and disc bolts, which transmit the torque from shaft to blades.

Damage

When a bird strikes the front end of a turbine engine several things can happen, depending on the size and weight of the bird, the speed of the aircraft, the rpm and type of engine, and the exact location of impact. Most single small birds are chopped up by the blades of the first rotor stage and pass through the interior of the engine without doing significant damage. With bigger birds the same thing may happen, but the blades of the first rotor stage may

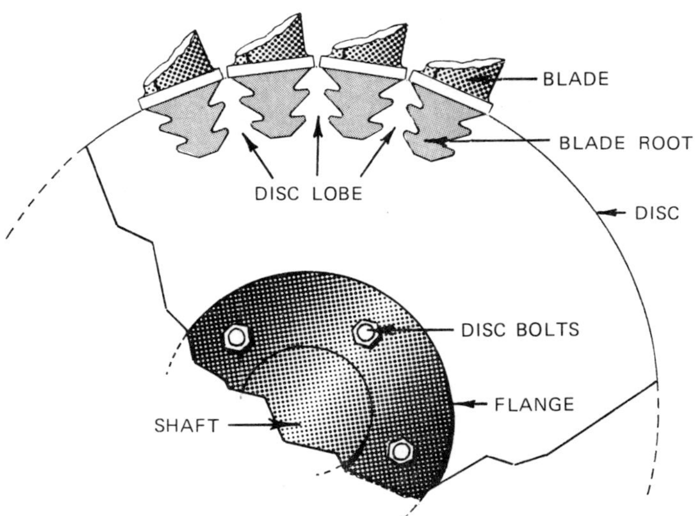

Fig. 3-4: Schematic view of a compressor disc showing how the rotor blades are attached.

be bent or deformed and would need to be replaced. In more serious cases, one or more blades break; the broken pieces can be thrown out either forward, sideways, or backward. The last case is the worst, because the broken piece of blade is likely to break more blades of the next stator and rotor stages. This chain reaction could destroy the compressor and the rest of the engine. If broken rotor blades of the first stage are thrown forward or sideways they may hit other parts of the aircraft, or they may be ingested by another engine of the same aircraft.

Apart from physical destruction, there is also the possibility that birds lodged on the inlet guide vanes or on the air intake screen will reduce the airflow into the engine. The Lockheed Electra crash near Boston (see Table 2-1) resulted from starlings that clogged the air intake screens of 3 of the 4 engines: the engines were thus unable to develop enough thrust due to the lack of air, and the aircraft reached stall speed and crashed (406).

Bird-proofing

To make turbine engines bird-proof, two approaches have been tried: (1) prevent birds from reaching the engine, and (2) change the design of the engine itself.

(1) KEEPING BIRDS AWAY FROM THE ENGINE. Attempts to keep birds away from turbine engines have included screens and grids in the air intake to intercept or deflect birds, and curved air intakes with a soft lining. A guard to "catch" the birds would have to be strong, otherwise the impact of a heavy bird at high aircraft speed would break it and send its pieces through the compressor. Not only the screen itself but also the way it is mounted in the air intake must be rigid. However, if the screen and its mounts are capable of withstanding the impact, then some other part has to absorb the kinetic energy of the bird. The impact shock load of a bird on an effective screen could be many tons and might tear the whole engine out of its mountings. Various bird screen systems have been produced, but all have their disadvantages (such as icing problems, weight penalty, reduction of engine performance, and the risk of airflow interruption). None has become standard equipment (10).

A more promising variant is an "inclined bird deflector grill" that deflects the birds away from the engine by removing them from the engine inlet through a special chute. This device was developed on behalf of the US Federal Aviation Administration for use in turboprops, and is shown in Fig. 3-5. The grill is retractable and would only be used at altitudes where birds were to be expected. The chute door consists of preloaded leaf springs, designed to remain closed during normal operations but to open by deflected impacts of different-size birds at low and high speeds (213). It is unknown to what extent this device has been incorporated in operational aircraft.

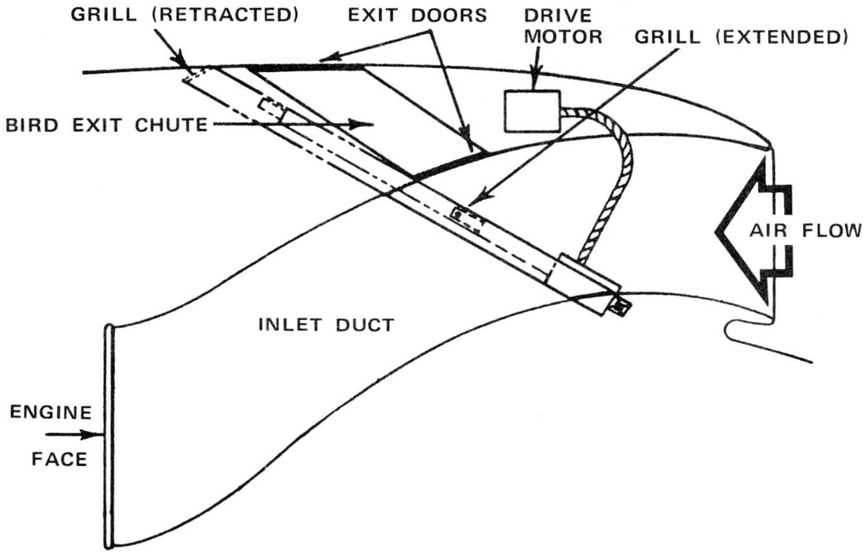

Fig. 3-5: An inclined bird-deflector grill proposed as a protection for turbopropeller engines (after 213).

Still another idea is a soft lining applied to curved air intakes, such as are present on the F-104 and the tail-mounted engine of the Trident. If a bird enters an offset intake, it may be dashed against the curved wall of the duct, damaging the metal wall so that pieces of the lining break off and also enter the engine. The suggestion is to make a soft lining so that the bird would become embedded in it, preventing damage to the engine proper. Tests with energy-absorbing materials including styrofoam, urethane foam, and aluminum honeycomb, however, showed that this approach held little promise (213).

In the case of a turboprop engine, the chances of engine damage from a bird strike are less than for a turbojet or turbofan engine. This is because of the relatively small air intake and the presence of a propeller rotating in front of it, which tends to chop up or knock away birds that might strike the air intake. Nevertheless, birds have entered the air intake of turboprop engines and caused damage; a Snowy Owl struck the engine of a Vanguard aircraft at Toronto International Airport and part of its body entered the intake, resulting in destruction of all compressor and turbine blades (48). This example also shows that the often suggested idea of having a propeller in front of the intake of a turbojet engine, to chop up any birds, would not always be effective. If a special low-drag rotor with closely spaced blades were to be used for this purpose, it would be as vulnerable as the first rotor stage of the compressor.

(2) CHANGES IN ENGINE DESIGN. The following components are normally affected by the ingestion of foreign objects (birds, hailstones, forgotten tools, etc.): (a) inlet guide vanes and rotor blades, (b) rotor discs and rotor disc bolts, (c) stator blades and compressor casing, and (d) nose bullets (117). A nose bullet is the streamlined cap of the engine's drive shaft.

(a) Inlet guide vanes and rotor blades The vane itself must be sufficiently strong to resist bird impact, and both ends have to be held securely so that neither end of the blade can be pushed back into the rotating blades of the first rotor stage.

If a bird hits one or more first-stage rotor blades, the blades have to move rearward a medium that is denser than air (i.e., the bird remains), and in doing so they are deflected forward. The deflected blade tips will come in contact with the trailing edge of the inlet guide vanes, unless sufficient space is left between vanes and rotor blades (117). The maximum deflection of the tip of a first-stage rotor blade after impact with a bird can be calculated. This distance would be the minimum clearance needed between the inlet guide vanes and the first rotor stage.

The greatest deflection of the first-stage rotor blades does not necessarily occur at the greatest speed of the aircraft. The aircraft speed at which greatest blade deflection occurs varies for different engines, depending on blade size, stagger angle of the stator blades, and engine (117). This implies that the bird strike equipment tests for rotor blades should be carried out at speeds where the greatest deflection is to be expected rather than at maximum speeds (9).

(b) Rotor discs and rotor disc bolts Bird impact loads on rotor blades are transmitted to the blade root and the reaction is

provided by the compressor disc lobes (Fig. 3-4). If a disc lobe were to fail, two complete rotor blades might be released. Thus, it is preferable to have a blade fail before a disc lobe does. This can be achieved by a careful choice of materials and engineering design, not described here.

The disc bolts of the first-stage rotor carry the impact torque, and they too can be made stronger by careful design (117).

(c) Stator blades and compressor casing Whereas the first-stage rotor blades may be exposed to maximum impact at relatively low speeds, the first-stage stator blades receive the maximum impact at the highest aircraft speeds.

In the design of bird-proof stator blades, attention is given to the blade itself and the way it is retained in the compressor casing. It is advantageous to join the tips of the stator blades at their inner ends with a ring, so that the impact loads are shared and the tips will not bend into the adjacent rotor blades (117).

(d) Nose bullets Stationary nose bullets can be made sufficiently strong to withstand a bird impact, or they can be designed to absorb the impact through collapse.

For rotating bullets only the first option is available, because bird impact will usually cause asymmetrical damage leading to serious imbalance of the engine.

Although most efforts have been directed at bird-proofing the first stages of rotor and stator blades, the whole engine must be considered. There has been a case where a bird passed through the low-pressure compressor at the engine's front end without causing damage, but then broke blades in the high-pressure compressor further downstream (9).

WINDSHIELDS

The following four factors are of paramount importance in the design of windshields: field of view, optical quality, structural integrity, and resistance to impact of hail and birds. In addition, there will be certain requirements relative to the windshield/airframe interface, methods of transferring loads, effects of aircraft configuration on the shape of the windshield, and the influence of aerodynamic drag and noise (27).

Broadly speaking, there are two types of windshield construction: flexible "bag-the-bird" and rigid "bounce-the-bird" windshields.

Until recently, most windshields in commercial jet transport aircraft were of the flexible type, consisting of a flat type of polyvinyl butyral (PVB) with glass on either side. The outer glass layer provides an abrasion-resistant surface to protect the PVB layer from being scratched by solid particles and windshield wipers. On its inner surface this outer glass layer carries an electrically conductive coating that de-ices the glass and also serves to keep the PVB warm.

The PVB layer provides resistance to impact. As its properties are altered to a great extent by temperature, it is essential to maintain a temperature near optimum for maximum strength when the outside temperature is low. Heating the PVB also relieves stresses resulting from large differences in expansion coefficients between glass and vinyl (321). The inner glass layer is designed to resist the pressure differential from the cabin to the outside.

The rigid type of windshield has recently been introduced for use in aircraft such as the L-1011 and B-747. These curved windshields use "stretched acrylic" with glass on either side, as shown in Fig. 3-6 (27). The glass layers are not bolted directly to the aircraft frame, the inner layer often being clamped to the acrylic via a Z-frame.

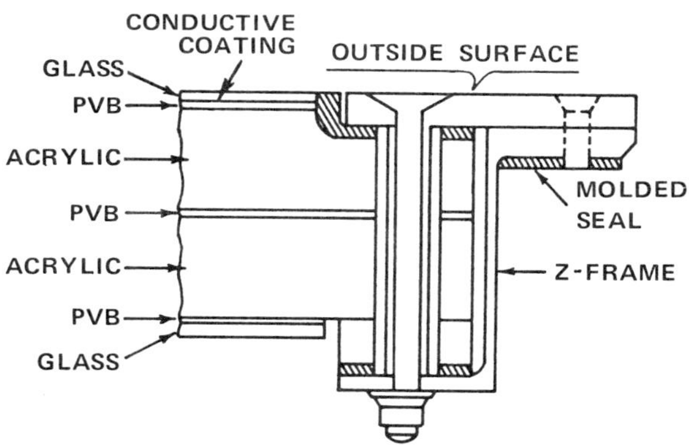

Fig. 3-6: Schematic cross-section of a curved acrylic rigid windshield on a modern airliner (after 27).

A major problem with both types of windshield is their relatively short durability because of deterioration of the optical quality due to delamination, coating burnouts, and abrasion by impact from hail and flying debris. As far as the bird problem is concerned, the results of strikes on windshields have varied from a smear of blood, to crazing or shattering of the windshield, to complete penetration of the bird into the cockpit causing injury, unconsciousness, and at least once, death of the pilot.

The designers of aircraft windshields are thus faced with a host of problems, of which bird impact is just one. The area of proper design boils down to the use of the right materials in the right configuration, at a competitive price tag. As with all bird-proofing, the problem is to absorb or reduce the force of the impact. When a highly rigid and strong windshield is used, the impact force is not absorbed by the windshield but transferred to the canopy framework, where it may well cause distortion. Although a distorted frame is obviously to be preferred to a smashed window pane, there remains the need to build some "give" into the system where damage is least likely to be disastrous.

The National Aeronautical Establishment of Canada published a bibliography on the hazards of foreign body impact on aircraft (174). The need for a theoretical model of bird impact forces on windshields was discussed in this report.

A Conference on Transparent Aircraft Enclosures was held in the US and its proceedings were published (414). Various materials such as polycarbonates and new plastics were described and their possible use in windshields was discussed. The many papers presented at this Conference were highly technical and are not further discussed here.

TAIL STRUCTURES

The tail assembly consists of the horizontal tail stabilizers (with elevators), the vertical fin, and the rudder (Fig. 2-3). The tail components are usually built from a stretched aluminum alloy over a frame of spars, struts, and ribs. A variety of designs exists, but their common goal is to provide strong yet light airfoils that can be controlled from the cockpit. The control lines are sometimes located in the leading edge of the airfoils, where they are likely to be damaged in a severe bird strike.

Although the tail components have had their share of bird strikes, most have caused only minor damage. The notable exception was the collision of a Viscount aircraft and Whistling Swans, which severely damaged the tail assembly and caused the aircraft to crash (see Table 2-1). It was probably this crash that led to the airworthiness requirement in the US that tail assemblies should remain operational after a strike by an 8-lb. bird.

In many cases, a head-on strike on the blunt leading edge of an airfoil results in a dent in the skin, or torn and crumpled metal, or an actual hole in the skin (Figs. 2-4 and 2-5). Sometimes the bird becomes lodged inside the structure, and in a violent collision the bird may pass through the spar, tearing a hole through the trailing edge as well.

Efforts to bird-proof the tail structure, and more particularly the horizontal stabilizer, have involved various types of modification to the leading edge, and included "doubler" plates, plastic-foam fillers, "splitter" plates, and leading edges of new design (2). Fig. 3-7 shows a cross-section of the leading edge of a horizontal stabilizer of a turboprop aircraft and two modified versions. Tests showed that the alterations were effective in preventing bird penetration, but the energy of impact was transferred to the structure behind the leading edge, causing it to buckle. The design changes were, however, effective in that they protected the rear spar, which is the main structural member of this type of airfoil construction. These and other designs have made it possible to meet the present airworthiness requirements for tail structures.

WINGS AND OTHER COMPONENTS

Although there are no bird strike requirements for wings, manufacturers have already tried to reduce damage from bird strikes. The wing airfoil is somewhat similar in shape, design, and construction to the horizontal stabilizer of the tail. Some improvement was obtained by reinforcing the leading edge and by placing "splitters" (to deflect the bird's body) between leading edge and spar. Attempts to reduce wing damage by filling internal spaces with a shock-absorbing material have not proved successful (250).

Strikes have been reported on pitot tubes, landing gear, extended landing lights, radio antennas, and on almost all forward-facing or projecting parts of an aircraft. With the exception of the pitot tube (a pressure-sensing device), damage to these components has been relatively unimportant, and no research and development work has been reported.

TEST PROCEDURES AND EQUIPMENT

New methods and equipment had to be designed which could be used to test bird-proofed components and to demonstrate to certifying authorities that new components meet the bird strike requirements. The testing techniques used have to simulate a bird strike.

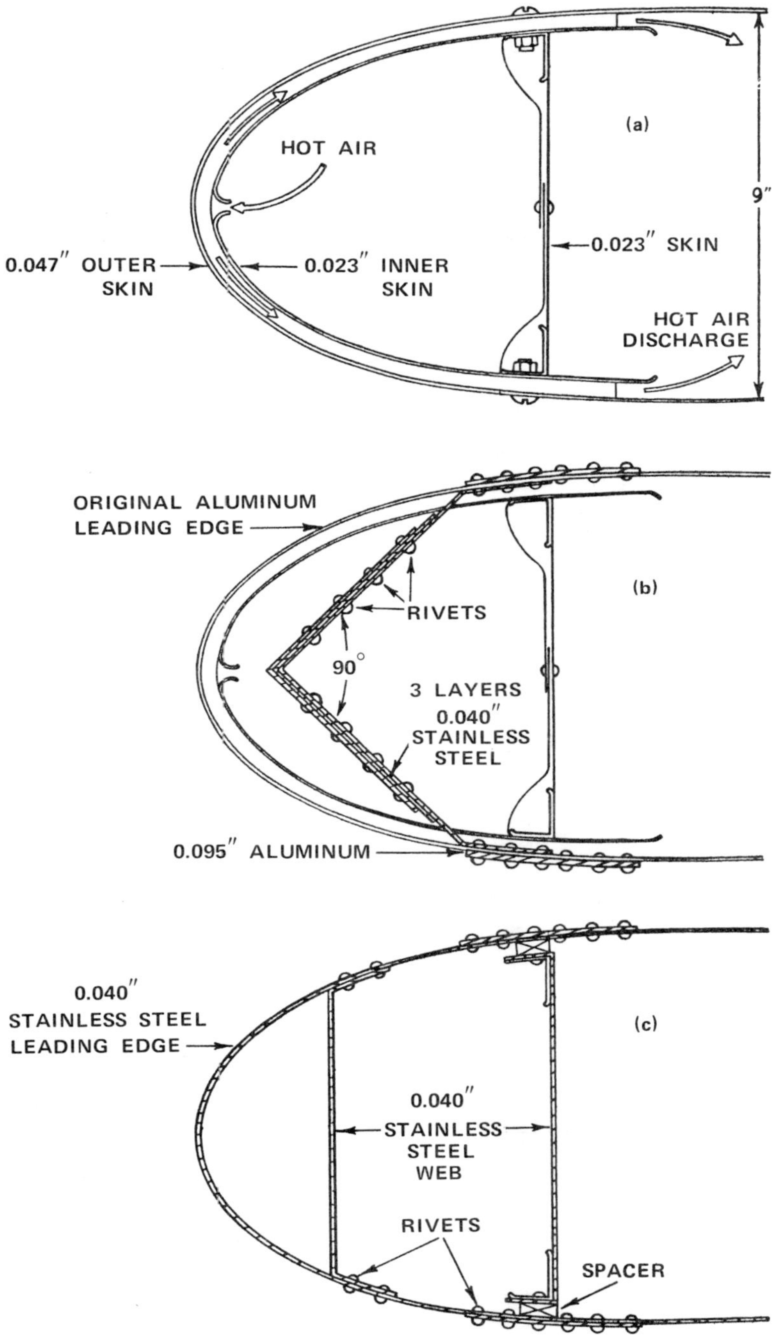
(a)
HOT AIR
9″
0.023″ SKIN
0.047″ OUTER SKIN
0.023″ INNER SKIN
HOT AIR DISCHARGE
ORIGINAL ALUMINUM LEADING EDGE
(b)
RIVETS
90°
3 LAYERS 0.040″ STAINLESS STEEL
0.095″ ALUMINUM
0.040″ STAINLESS STEEL LEADING EDGE
(c)
0.040″ STAINLESS STEEL WEB
RIVETS
SPACER

From basic considerations it was thought that for true simulation the target (e.g., a windshield) should be propelled into a stationary bird, rather than vice versa (153). This approach involves the use of equipment such as a rocket sled, on which is mounted the aircraft component being tested. In the other approach a specially designed gun is employed to propel a bird against an aircraft component. From detailed comparison of these two approaches it was concluded that the state of the bird during impact is adequately simulated by either sled or gun experiment. The behaviour of the aircraft component can also be properly simulated by either technique, provided that special measures are taken to reproduce in-flight conditions (152). For example, windshields may have to be cooled to represent upper air temperatures.

A rocket sled travels on several miles of rails and is propelled by rockets to provide extremely high speeds. The test birds are

Fig. 3-8: Test birds suspended over the tracks of a rocket sled that is used in the United States to simulate bird strikes on aircraft components.

Fig. 3-7: The leading edge of the horizontal stabilizer on a turbopropeller aircraft is shown schematically in cross-section at the top (a), with two designs below (b and c) modified to withstand bird impact (after 2).

suspended above the track (Fig. 3-8). A speed-measuring device, a motion-picture camera, and a dynamometer to record the impact force are required. Rocket sleds are costly to install and operate, and sometimes have a bird strike problem of their own. The rocket sled at Holloman AFB, New Mexico, has had several collisions with birds, costing an average $25,000 per strike (442). In one test a 3/8 inch-thick steel cover plate on an instrument package struck a Horned Lark, which deeply dented the plate and destroyed the instrument (355).

The Impact Simulator Gun is much cheaper to build and use and has the advantage that impact tests can be carried out with more convenience, better control of the environmental conditions, and greater ease of recording details of the impact (152). Such bird guns are used in Canada, France, Germany, the UK and the US. Fig. 3-9 shows the equipment used by the National Aeronautical Establishment (NAE) in Canada. It is basically a big air gun consisting of a large reservoir of highly compressed air, a rupture-diaphragm, a long barrel, and some associated gear. The equipment is capable of propelling a 4-lb. bird at speeds of up to 818 mph (61). High-speed motion-picture cameras, special lighting equipment, and other devices are used to record details of the impact.

To avoid the spatter of blood and tissue around the impact areas, use of synthetic "birds" has been attempted. However, results with these mixtures of grease and fiber are not fully accepted as being representative of the effect of the impact by a real bird. The test installation of the NAE has been equipped with wheeled baffles that can be taken outside the test building for cleansing.

In standard tests, a 4-lb. bird has been generally adopted. Usually chickens are used, and one agency specifies that these must be "barnyard fowls" because their body structures may be stronger than cage-reared chickens and so provide a more rigorous test. It is the practice to enclose the dead bird in a cloth or plastic sack to prevent it from disintegrating from air pressure during its passage from gun to target. The bird and sack are also enclosed in a sabot casing for discharge from the gun barrel, the sabot remaining behind at the muzzle.

Other tests are designed to simulate the impact of several smaller birds within one second. In these experiments a multi-barrel bird gun is used.

The development and testing of bird-proofed aircraft components, and particularly of turbine engines, is a costly undertaking involving aeronautical experts, costly equipment, and occasion-

Fig. 3-9: (top) A bird impact simulator gun installed at Ottawa, used for testing aircraft components. (bottom left) Experimental windshield smashed by the impact of the test bird. (bottom right) The dummy's face and chest are covered with debris of the test bird and shattered windshield.

ally the write-off of the structure under test. Although certifying tests for engines have to be carried out on complete engines, in the development and testing of rotor blades it is not necessary to use a whole engine. Fig. 3-10 shows a test rig for single blades used in the UK (410).

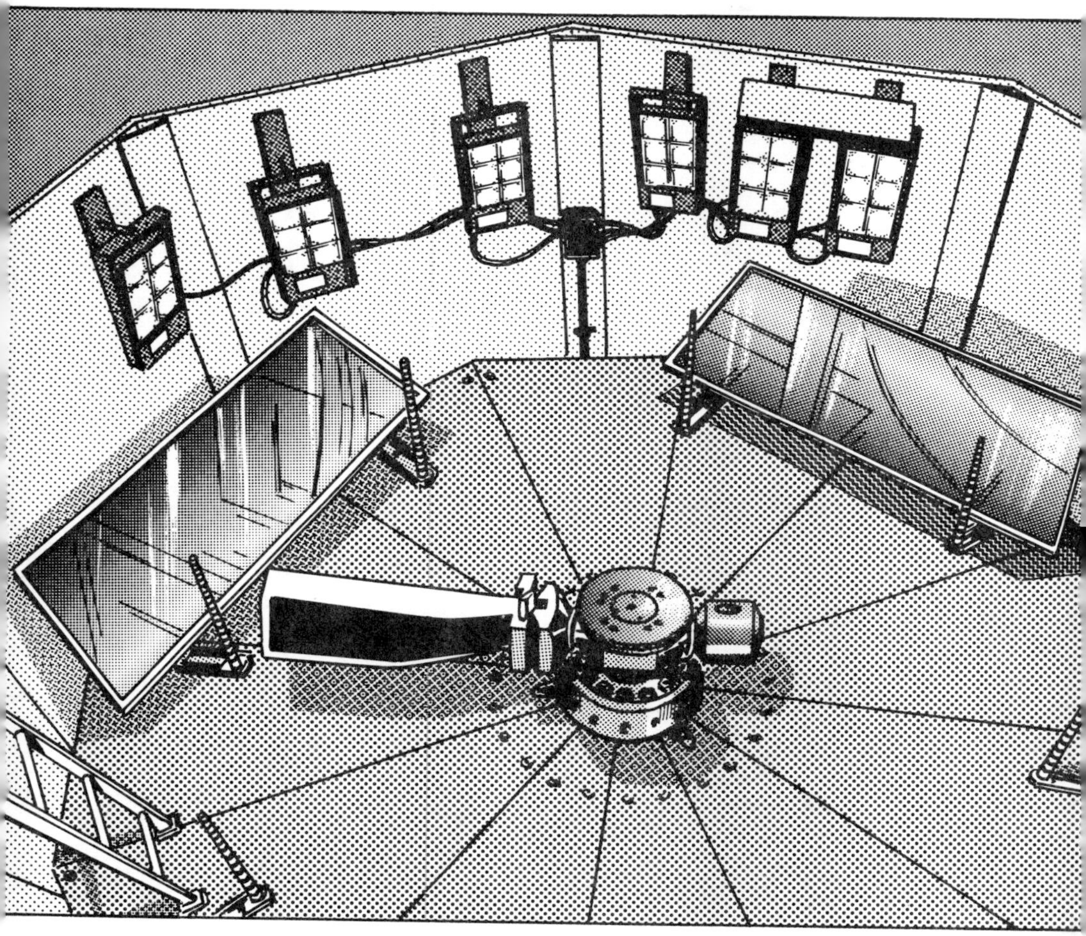

Fig. 3-10: Apparatus for testing the effect of bird impact on a single rotating compressor blade (after 410).

CONCLUSIONS

In recent years turbine engines, windshields, and tail structures have been developed that meet current airworthiness requirements. This represents a great reduction in the bird strike hazard. The chances of a disaster resulting directly from the impact of a bird strike on a modern, fully certified aircraft would now seem to be extremely small. Nevertheless, as discussed in Chapter 2, several close calls have occurred even on modern certified aircraft, and further research and development in this area is fully justified. In addition, there may be a need for stricter bird strike requirements for certain components or aircraft types.

There remains also the risk that a bird strike will cause a potentially disastrous accident in an indirect way. These "multiple causation accidents," which often include some misjudgment or error by the pilot, are mentioned in Chapter 2.

Even if bird-proofing efforts are greatly increased, it is not likely there will ever be a completely bird-proof aircraft that will also be able to fly. Thus there is good reason to look into other methods to reduce the bird hazard, while continuing bird-proofing work at the same time. The next chapter describes the search — unsuccessful to date — for an on-board device that would get rid of birds in front of the aircraft.

Apparatus used in testing the effect of microwave radiation on chickens.

Chapter four: Get out of my way!

THE SEARCH FOR ON-BOARD EQUIPMENT TO DISPERSE BIRDS

This chapter reviews some of the attempts made to develop a device that could be installed in an aircraft to clear birds from its flight path.

If on-board equipment could cause birds to turn quickly out of the flight path of an approaching aircraft, bird strikes could be avoided. Such a device would be required to:

- be effective on birds of all species under all conditions,
- have no harmful effect on humans,
- have no adverse effect on aircraft performance,
- be readily installed, operated and maintained,
- be of reasonable cost and weight, especially when installed in existing aircraft.

However, it appears unlikely that any device will be available in the near future to meet even one of the above requirements. For various reasons, touched on below, results of experimental work are not encouraging. Three forms of electromagnetic energy (light, laser, and microwave) have been considered for the development of an airborne bird dispersal unit.

ON-BOARD LIGHTS

In Chapter 1 it was noted that some bird species, especially waterfowl, actively try to avoid approaching aircraft. Yet waterfowl have been involved in strikes, some of the strikes occurring at night. It is not known whether birds are hit more often at night than during the day, assuming no differences in bird density and in aircraft density and speeds. It is also not known whether birds that avoid aircraft do so because of the noise, the appearance, the speed, or the general "unnaturalness" of aircraft. At any rate, it seems logical to assume that if aircraft are made easier to detect, birds would have more time to get out of the way. It is likely that in many strikes the birds have seen approaching aircraft too late; or if they see them they may be unable to judge the aircraft's speed and direction, and thus fail to realize they are on a collision course. The presence of strong lights on the aircraft would give the birds more time to detect it and to anticipate its flight path.

Large aircraft usually carry the following lights: relatively weak rotational beacons on top and underneath, weak coloured lights at the wing tips, powerful landing lights in the wing, a white taillight, and a powerful taxi light in the nose-wheel unit. Wing-mounted landing lights are often recessed in the leading edge of the wings and covered with a protective material such as perspex, or they are retracted into the wing when not in use. With the latter design, landing lights can only be employed well below cruising speeds.

Despite the presence of the rotational beacons and other lights, pilots often have difficulty spotting other aircraft in the sky. It has been urged that aircraft be fitted with powerful flashing or strobe lights, to increase their visibility both day and night and to reduce the number of midair collisions and near-misses (105). Thus strobe lights would serve the dual purpose of providing easier detection of the aircraft by man and bird.

Little is known about the behaviour of birds with respect to flashing lights in general, and nothing is published regarding their reactions towards aircraft-mounted strobes. A study of caged pigeons subjected to treatment with high-intensity flashing lights showed that exposure to a strobe acts as a stress factor: the birds ate less, routine activities were disturbed, and their adrenal glands increased in weight (68).

Another study with caged birds showed that Common Starlings kept in a dark room responded to pulsed light by increased activity (i.e., more hopping on the perch). The higher the pulse frequency (a maximum of 400 pulses per minute was used) the greater the response of the bird. Somewhat similar results were obtained when the starling tests were repeated in a lighted room. However, when Mallards were treated the same way, they did not give any significant response (their level of excitement or irritation was estimated by measuring their heart beat rate). After they became accustomed to the light, they soon lost what little response was seen (279).

There is no proof, and little evidence, that the use of airborne strobe lights reduces the number of bird strikes. Two "comparable airlines flying similar equipment to identical airports" were reported to suffer different amounts of bird strike damage, with the airline using strobe lights paying a much smaller repair bill (105). Statistically, these results may not be relevant because it is doubtful that they are strictly comparable (e.g., one company may have had more nighttime operations during a few nights with heavy migration, or may have serviced more airports along the route). Tests with USAF fighter aircraft, half of them fitted with a strobe, half without, but all flying similar routes over similar terrain were inconclusive because only a few strikes were reported for both groups of aircraft (123).

Based on a few incidental observations, especially of geese that actively avoided approaching aircraft with landing lights on, Air Canada has encouraged its pilots to have "landing lights on below 10,000 feet." Statistics for strikes that occurred after the introduction of this policy seem to indicate a decrease in numbers (46). The "lights on below 10,000 feet" rule has also been adopted in Sweden (387), and in the US the FAA recommends that landing lights be turned on during operation: (1) within ten miles of any airport, day or night, (2) under reduced visibility, as in haze, dusk, smog, etc., (3) under special Visual Flight Rules (VFR) conditions where such operation is allowed, (4) where flocks of birds may be expected (in coastal areas, near swamps, lakes, refuse dumps, bird sanctuaries, migratory flyways, etc.). In addition, pilots are asked to use anti-collision lights (rotating beacons) whenever the engines are running, day or night (17).

Even though the extensive use of landing and anti-collision lights has now become routine procedure in some areas, there remains the need to develop a much stronger strobe light that would help both pilots and birds. A great deal more research has to be done on the effect of strobe lights of different colours, intensities, and flashing rates, both on people (annoyance, dizziness) and on captive and free-flying birds. On nights with poor visibility, migrating birds are known to crash into lighthouses, lighted TV towers, and lighted tall buildings. It may well be that on such nights migrating birds would be attracted rather than repelled by aircraft-borne strobe lights. This would mean that on nights with poor visibility the "bird beacons" would have to be switched off.

ON-BOARD LASERS

The word "laser" is an abbreviation for "Light Amplification by Stimulated Emission of Radiation." Laser devices produce either pulses or a continuous beam of intense light.

Early experiments in the US with lasers aimed at the heads of caged gulls showed that although their eyes were physically damaged the birds were not distressed or alarmed. Thus it was decided to await the time when more powerful lasers would become available before carrying out further experiments (357).

At a UK conference in 1967 it was suggested that the possibilities of medium-power gas lasers be investigated in the visual range operating at flicker frequencies corresponding to central nervous frequencies. If these lasers were not effective, then higher-power lasers "could burn off or wither the flight feathers of birds in the immediate track of climbing or level-flying craft." Such powerful lasers would have to work in the infrared region of the light spectrum, to avoid damage to crew members behind windows of distant aircraft (348).

Further research on the feasibility of laser bird control commissioned by the USAF was reported in 1972 (279) and 1973 (280). In this study Common Starlings, Mallards, and Herring Gulls were exposed to lasers of different intensities and pulse frequencies, and their behaviour and physiological responses were monitored. The birds were irradiated with both a low-intensity strobe light pulsing at frequencies from 100 to 1,000 per minute and an argon laser capable of emitting either pulsed or continuous light of different wavelengths at intensities of 20 milliwatts to 4 watts. The intensity of laser light striking the bird could be varied further by changing the diameter of the laser beam.

A concentrated beam (0.08 in. in diameter) at power above 0.5 watt caused an avoidance response in the three species tested. It was the only light source to which the birds did not become accustomed. The initial response to lower intensity lasers (beam expanded from 2 to 14 in.) was increased excitability (increased perch hopping and heart rate), but gulls and starlings soon became habituated. Ducks were more sensitive than gulls or starlings to the expanded beam. They had elevated heart rate, and avoided the laser beam expanded to 6-inch diameter at least 50 per cent of the time (279).

Because concentrated laser beams that irritate birds would also irritate or injure people, use of concentrated lasers on-board aircraft cannot be considered, quite apart from the technical problems of installing and operating them. However, an expanded laser beam of low intensity, with greater effective range than normal landing lights, might be used as a superior anti-collision beacon, giving birds and pilots more time to avoid aircraft.

As with strobe lights, more research and development is clearly required to determine the usefulness, safety, technical feasibility, and costs of an on-board laser warning device.

ON-BOARD MICROWAVES

"Microwave" is the term used for electromagnetic waves with wavelengths of 0.1 cm to 1 m. Microwaves are used in different types of radar and in microwave ovens.

As radars have become an extremely important tool in bird migration studies, it is essential to know whether radar affects the behaviour of flying birds. In his book *Radar Ornithology*, Eastwood devoted a chapter to this problem. After reviewing conflicting reports and conducting special tests, he concluded that "radar is, indeed a legitimate tool to use in ornithological research and does not perturb the birds whose presence it detects" (138). This statement referred to existing radars operating in a normal way. Radars with different wavelengths, power level, or pulse repetition frequency might have some influence on birds either as a "thermal effect" or perhaps by interfering with their navigation system.

The Canadian Committee on Bird Hazards to Aircraft decided to support further investigations on the effect of radio waves on birds. The program included studies of the effects of microwaves on pigeons, gulls, and chickens. The observed effects included avoidance reactions, distress, and muscular disturbance (390-393), but in all tests the birds were close to the microwave source.

There is no doubt that microwaves affect birds at short range. However, to use an airborne microwave source to "paralyze" a bird so that it would drop clear of the flight path, the minimum effective range for an aircraft flying at 180 mph would be 0.5 mile, assuming the bird's reaction time to be 10 seconds. To be effective, the microwave source would have to be extremely powerful. At present, use of an on-board microwave generator does not appear to be possible for this reason (222).

At an airport, the power required to disperse birds would likely be so high that a microwave device could not be used because of the unacceptably high radiation level to people in the area (222). For these reasons, the work has been abandoned by the Canadian Committee.

CONCLUSIONS

There appears to be no effective airborne device to disperse birds from the path of an aircraft. Further studies on the potential of on-board strobe lights (employing either normal or laser light) to alert and repel birds seem to hold the greatest promise at present. It is, however, unlikely that such lights would be effective under all conditions, and it may well be that other factors (distraction of the crew, etc.) may preclude their operational use.

Raptors such as this Snowy Owl and Great Horned Owl trapped at Toronto International Airport can be a hazard while hunting near runways. Raptors captured at the Toronto Airport are banded and released some 20 miles away (photo: Toronto Globe and Mail*).*

Chapter five: Airports are not for the birds

PREVENTION OF BIRD STRIKES AT AIRPORTS

This chapter reviews some methods that can be used to prevent bird strikes at or over airports. The aim is to get rid of birds at and around airfields by scaring, trapping or killing the birds, or by making the airport itself less attractive to them.

At any airport in the northern hemisphere it is likely that breeding birds will be found in spring and summer, migrating transients in spring and fall, and visitors during the winter. In addition, a few species may be year-round residents. Some of these species, although not numerous, may be strike-prone, whereas others that occur in large numbers at an airfield may nevertheless rarely be struck by aircraft. Local breeding birds become well accustomed to air traffic and are not likely to be involved in collisions. Their offspring, however, have to learn to stay away from aircraft, and many young birds are struck each summer. The local breeders are hard to fool with scaring devices to which they gradually become accustomed, whereas migrant birds, pausing at the airfield, may respond much more readily to scaring equipment.

If scaring does not work and if the species involved is not numerous yet is a serious hazard (e.g., Snowy Owls, which are winter visitors to Canadian airfields where they hunt over runways and taxiways), it is possible to live-trap and remove them. Some of these birds, however, come back to the airfield because the attraction remains. If they do not return, other birds of the same species may move into the vacant habitat.

Trapping and killing of large numbers of birds usually serve no useful purpose, because any unoccupied area will soon be filled by birds from neighbouring areas.

When it is difficult to get rid of birds from an airfield, the airport itself can be altered to make it less attractive to them. Once the features that attract birds to the airport are known, they can be removed or changed to reduce their attractiveness. This approach is called habitat manipulation, because it involves alterations of the habitat (or living space) of the birds. Changes in the airport environment are useful in removing the more obvious bird attractions, but it will be impossible to change the airport in such a way that it will attract no birds at all.

Ideally, when a new airport is planned, a site with a low biological capability should be chosen so as to minimize the future bird problem. An "anti-bird" attitude should be maintained throughout the design, construction, and operation of the new airport. If, despite these precautions, bird problems occur, the birds should be frightened away. If that fails, they can be trapped and released far away from the airport, and if that fails too, they may be killed. Killing is a last resort, and an admission that usual control and management techniques were inadequate.

In practice, however, the situation is often completely different, with a reverse sequence of events. Sudden bird problems at existing airfields need immediate answers. An extensive change in the airport habitat (the "ounce of prevention") is usually costly and time-consuming. Also, the decision where to build a new airport is usually based on political and socio-economic arguments rather than on biological factors. Although the ideal approach is probably cheaper in the long run, often a few costly bird strikes must occur before large-scale habitat manipulation will be undertaken.

In summary, current methods of bird control are far from perfect; no cheap, simple, yet effective methods for scaring birds have been developed; and there is a limit to what can be achieved by habitat manipulation.

BIRD OBSERVATION METHODS

Before any effective scaring or trapping can be undertaken, the whereabouts of the birds on the airfield must be known. During the day, bird observations can be made from the control tower or a patrol vehicle (with binoculars or spotting telescope) or with television cameras. Closed-circuit TV located on the field (with remote-control cameras installed at critical areas) allows bird observations to be made from dawn to dusk. Such observations are impeded by reduced visibility and are impossible on dark nights.

Since many birds can be present at airfields during the night, either resting, feeding or flying about, nighttime observations may be required. They can be made with searchlights, night viewing devices, or a specially designed bird-detection radar.

NIGHTTIME OBSERVATIONS WITH A SEARCHLIGHT

In New Zealand a strong spotlight, carried by hand and powered by a wet-cell battery pack, was used to locate and identify birds at Auckland's International Airport (345). Strong searchlights that can be rotated and elevated, such as can be found on top of fire trucks, are useful as well, but they may affect the birds' behaviour. In addition, on soggy airfields the truck would be restricted to paved surfaces and their intense lights may have a bad effect on the night vision of pilots using the runways and taxiways.

NIGHT VIEWING DEVICES

The Canadian Armed Forces made a study of available literature on night vision equipment and concluded that three types could be used to observe birds at night: (1) direct-view image intensifiers, (2) low light level television (LLLTV), and (3) active infrared viewers. All three types of equipment use image intensification, whereby minute amounts of light are "amplified" electronically. This is accomplished by one or more image intensifier tubes. These consist of a light-sensitive cathode (emitting electrons where struck by light), an electron focusing system, and a phosphor anode (that becomes fluorescent when struck by electrons).

(1) Direct-view image intensifiers are the simplest, lightest and cheapest of the three types of equipment. Sometimes called "starlight scopes," they are basically a telescope with image intensifier tubes inserted between the objective and ocular lenses. They are self-contained units using dry-cell batteries as power source. Weighing between 3 and 17 lb., they can be handheld or mounted on a tripod (104). Many of these viewing devices have been used only by the military and information was usually not easily available. However, recently two night viewing devices were tested for viewing birds at night, and several studies of nocturnal birds and mammals now appear feasible (288).

(2) LLLTV consists of an objective lens, an image intensifier tube, and a television camera. One of the advantages of this technique is that the TV screen can be recorded on videotape or film for later analysis. LLLTV's are designed for remote viewing. At an airfield an LLLTV could be mounted to survey a critical area, and the display unit could be installed in the control tower. Remote control would allow scanning in the horizontal and vertical plane, and focusing and changing of lenses.

(3) Infrared (IR) night viewing instruments are similar to direct-view devices, but the cathode of the first stage of the image intensifier is sensitive to the IR portion of the light spectrum rather than to the visible part. Commercial instruments require that the target be illuminated with an IR source, and the range of visibility depends on the power of the IR source.

Night viewing instruments are not in general use for bird detection at airfields. However, for airports with many aircraft movements at night and with a record of nighttime bird strikes, the use of LLLTV or other night viewing devices should be seriously considered.

DAY AND NIGHT OBSERVATIONS BY RADAR

Radar can detect not only airborne targets (such as aircraft, birds, and insects), but also targets on the ground provided that they stick up sufficiently high above the surface. Tests were made of a low-cost radar that would display a map of the airfield and show the presence and magnitude of bird groups on the runways and in shorter vegetation, under all weather conditions except heavy rain or snow. A high-resolution radar operating at a wavelength of 3 cm (such as marine radar) with a modified antenna at a height of 60-90 feet above the ground would meet these requirements. This ABDE (Airport Bird Detection Equipment) radar could also be used to locate disabled aircraft, and to control emergency vehicles (350).

Once the presence of birds or other undesired objects is detected, a bird-scaring patrol should frighten the birds away. As the airport would need to have a bird-scaring team anyhow, it might be cheaper to equip this patrol with a night viewing device, and to instruct it to search actively for birds as well. The purchase, installation, operation, and maintenance of an ABDE radar would have to be weighed against the cost of having a larger or better-equipped patrol unit. Operational use of ABDE's has not been reported.

BIRD DISPERSAL METHODS

All birds-scaring methods involve some form of "behavioral manipulation," that is, the birds are provided with a stimulus that changes their behaviour (e.g., they stop eating worms on the runway and fly away). In general, the stimulus must be perceived by the birds via their senses. It is, of course, also possible to expose the bird to physical influences that affect their body in a direct, extrasensory way (i.e., not via the senses). Microwaves, as discussed in Chapter 4, could fall in the latter category. In this chapter we will only deal with stimuli perceived via the senses.

As described in Chapter 1, most birds have excellent sight, reasonable hearing, rather poor smell and taste, and little is known about their sense of touch.

The sense of taste in birds has not been used in bird clearance operations. Theoretically, an airfield could be sprayed with a bad-tasting chemical (a so-called gustatory repellent) that would be ingested by soil invertebrates, which would thus become unpalatable to birds. Birds frequenting the airfield to feed on these soil invertebrates would probably leave the area in search of unpolluted food, but other birds would be unaffected (443).

The only olfactory repellent (effective through the sense of smell) that has been tested on airfields is naphthalene, which was applied as "moth balls" to several airfields in the UK. Reported results were contradictory, and when tests were repeated no conclusive evidence was obtained (443).

The sense of touch offers some opportunities to get rid of birds in and on buildings, such as hangars. Sticky material is applied with a caulking gun to window sills, ledges, rafters, and other parts of the building that birds like to perch on. Birds dislike sitting on this soft and sticky substance and will avoid such places. Eventually the material dries out, becomes dusty, and loses its repellent properties.

Electrical shocks are also perceived via the sense of touch and can be used to scare birds. Electrical wires are strung over the birds' perches and when they alight on them, or otherwise touch the wires, they get a nasty "tickle shock" of high voltage but low amperage. The shock does not do any permanent harm but has a good repellent effect. The technique is costly, but it has been applied with good success to prevent the fouling of statues and official buildings. It has not been used at airports.

Another approach to get rid of birds by playing on their sense of touch was the idea of luring gulls away from the runway by providing them with electrically heated "hot pads" near the perimeter of the airfield. This technique was based on the assumption that gulls like to lounge on the tarmac because it is warmer than the surrounding grass, but the method was apparently unsuccessful (207). Similar laboratory experiments with gulls in

Canada were also unable to confirm this assumption (344).

Apart from the few examples mentioned, all efforts to frighten birds at airports operate on the birds' faculties of sight and hearing. Scaring methods can be categorized as follows:

(1) visual scaring: bird corpses, bird models,
(2) acoustical scaring: ultrasonic sounds, non-natural sounds, natural sounds, and synthetic sounds,
(3) combined visual and acoustical scaring: pyrotechnics, birds of prey, remote-control model aircraft.

These categories are rather artificial. A shellcracker shot from a patrol vehicle to explode on the ground among the birds is basically an acoustical device, but the flash of light or puff of smoke during the explosion may have a frightening effect as well. Similarly, live hawks are mainly a visual means to scare birds, but their calls may add to the frightening effect.

Birds quickly become accustomed to frightening stimuli that are no real danger. To prevent or delay this habituation, the stimuli have to be changed often or reinforced by other means.

Birds were pests in agriculture before they became a nuisance in aviation. Most techniques to scare birds were developed to reduce damage to crops. Many governments are committed to protect birds as well as farmers' interests. Bird-control methods are thus often developed and tested by government agencies to aid agriculture on the one hand, and to prevent unnecessary destruction of birds on the other. On some occasions, governments supervise or participate in bird-control operations. In Canada, for instance, most provinces employ pest control officers, who advise and assist in cases where man's interests are damaged by wildlife.

The literature on bird control is scattered, but useful articles are published in the *Journal of Wildlife Management* and in *Pest Control.* The published proceedings of meetings of the International Union for Applied Ornithology, the annual Bowling Green Bird Control Seminars, and the California Pest Control Seminars are other important sources of information. In 1967 a symposium was held in England on bird problems in aviation and agriculture, and proceedings were published (301). Most books on pest control are largely devoted to insect problems, but some deal with birds as well (246, 418).

VISUAL SCARING METHODS

Birds are hard to fool. That is the main lesson man has learned from his many attempts to frighten birds by putting up scarecrows, or contemporary counterparts.

In Holland temporarily successful results were obtained by exhibiting stuffed gulls in abnormal positions (Fig. 5-1). This method was abandoned because it could only be used during the daytime, the stuffed birds had to be moved regularly to prevent habituation, and rain-soaked specimens soon lost all effectiveness as a repellent (126). Similar problems arose when dead gulls were used at Schiphol Airport. There the birds were dangling from posts and swinging in the wind, and again, the results soon were negligible.

At Schiphol further experiments are underway using styrofoam cutouts of gulls. When laid out on the grass their effect was nil, but when hanging from a post they appeared to have some repellency under certain wind conditions (212). UK researchers are also planning to investigate the possibilities of gull cutouts (75).

Fig. 5-1: Stuffed gulls in abnormal positions have been used at airbases in Holland to drive gulls away.

Experiments with bird decoys were carried out in the UK. Hunters and bird catchers often assert that decoys should be placed in a natural position, with their heads more or less into the wind. Decoys in unnatural positions will, they maintain, repel rather than attract birds. Stuffed models of Herring Gulls and Lesser Black-backed Gulls, in both natural and unnatural positions, did not appear to have any influence on where visiting gulls would land (443).

Results of preliminary experiments with mounted gulls in the US were promising and additional work is planned (384).

The most encouraging results were reported from New Zealand, where corpses of Black-backed Gulls (preserved with formalin) nailed on boards, eliminated lounging gull flocks. Usually the effects lasted for as long as the corpse was reasonably tidy, about three months (345). A further development was the use of polystyrene models of board-mounted birds, which were found to be equally effective and more durable. Good results were obtained with these models at New Zealand's Wellington Airport, where two traditional roosts were permanently eliminated. This success may have been partly due to the presence of suitable sites for roosting nearby. Results at Whangarei Airport were less impressive, because there was a dearth of alternative roosting sites (especially at high tide) and because the airfield also served as a night roost, when the models had no value (88).

Regarding the use of light as a means of visual scaring, Chapter 4 mentioned two studies which showed that some species become irritated or excited when exposed to strobing or flashing lights. A strong strobe light installed on an aircraft would not necessarily scare flying birds away but might give them more time to avoid the aircraft.

The use of fixed strobe lights to get rid of birds at airports is restricted to the inside of buildings. On the airfield, birds would soon become completely used to the flashing lights, which would probably be more annoying or confusing to pilots than to birds. Commercially available strobe lights have been recommended by the manufacturers for use against pest birds inside buildings such as storage sheds, food processing plants, barns, etc. Bird-repelling strobe lights, installed in two hangars at Montreal International Airport, produced good results in one hangar but were not effective in the other. It appeared that one hangar had many holes that provided easy entrance (and exit) for the birds. The other hangar offered no access for the birds other than the big hangar doors; it was in this hangar that the strobe lights were effective (48). Revolving lights, tested at Pueblo, Colorado, as a repellent against House Sparrows nesting in a hangar, were partly successful in reducing the "dirtying" problem caused by the sparrows. Subsequently this problem was virtually eliminated by use of a bird-scaring device known as a carbide exploder (356).

The use of lasers to frighten birds from airfields does not seem feasible at present. Tests with caged birds in the US (279) showed that in order to move birds, the laser beam has to be of such a high intensity that it would be highly hazardous for man. In a review of a study in the USSR on the effects of laser light on pigeon eyes, it was concluded that "laser beams are too dangerous for animals and humans for use in repelling birds from airports" (244).

Even if a six-inch-high black metal shield were installed around the perimeter of the airfield to trap the beam and if infrared light were used to reduce possible damage to the human eye (as was suggested in a US study), there would still be ample opportunity for regrettable errors. Also, whether the laser system were to be installed in a truck, or positioned in the field so that it could scan a large area, it would require specially trained personnel to operate and maintain it.

A number of other visual scaring methods have been tried, but were found unsuccessful or impractical. These included: coloured smoke, blue balloons, a purplish coloration of the grass, a person in full view of the birds flapping his arms in mock wing-beats with a frequency of about 25 beats per minute (443,445), and the spreading of large panels of brightly coloured fabric around the grass area which the birds occupied (197). In Canada, rubber owls placed in the rafters of hangars were ineffective in driving off sparrows and starlings.

ACOUSTICAL SCARING METHODS

Four types of sound have been used to get rid of birds: (a) ultrasonic sounds (i.e., above the frequency of human hearing, over 20,000 Hz), (b) non-natural sounds (claxons, bangers, etc.), (c) natural sounds (e.g., taped distress calls of birds), and (d) synthetic sounds (experimentally developed sounds).

(a) Ultrasonic sounds

Several investigators agree that birds do not respond to ultrasonic sounds. As one scientist put it: "In fact most birds cannot hear them, and if they could, these high-pitched narrow-beam sounds would mean little to them" (62). And, even if it should mean something to the birds, the use of ultrasonics would probably be impractical on large areas, such as an airfield, because of the large amount of power required due to the rapidity with which high-frequency sound loses strength with distance (443).

Incidentally, the use of infrasound (i.e., below the frequency of human hearing, about 20 Hz) is not being considered for bird dispersal because the effects of airborne infrasound, at least in man, become significant only at high intensities of over 130 dB (248).

(b) Non-natural sounds

Numerous devices (claxon horns, bells, rattles, chimes, sirens, exploders, drums etc.) have been used to produce sounds to frighten birds (156).

Loud noises in themselves do not seem to bother birds. Birds have been found breeding along noisy runways, and a pheasant was seen feeding at about 10 yards from a runway from which two F-104 Starfighters took off simultaneously, both using their afterburners. Whereas the human observer needed to protect his ears, the pheasant did not once stop pecking and did not seem to be disturbed.

Claxon horns, tested at a British airbase, were unable to frighten the birds permanently (443). Acetylene cannons (with bangs of 110 dB at 3 feet from the source, and with one explosion every 10 minutes) were also tested at a British airbase, but the results were inconclusive, although they showed that the cannons had better results when used in combination with silhouettes of figures with guns and when the location of the two types of stimuli was changed frequently (443).

Experiments to test the idea of using high-intensity sound for repelling Mallards and Pintails from cereal crops showed that the birds were driven away by the high-intensity sounds, but soon returned (398). Further tests were undertaken to find out how much noise Ring-billed Gulls can stand. The acoustic irritation threshold was about 85 dB for sounds of 400-500 Hz. Because such high sound levels are needed to produce a painful experience for gulls, it was concluded that there was little hope of controlling gull movements at an airport through the use of high-intensity sounds (399).

In farming operations, gas cannons have often been used, with varying degrees of success. Satisfactory results are usually only obtained under special conditions. During their fall migration across the Canadian prairies, cranes and waterfowl feed on grain near their staging areas. They often do a lot of damage to the crops in a short but critical period. These migrant flocks can be scared away by prolonged systematic harassment with gas cannons, provided that they are allowed to feed undisturbed in certain fields nearby, where they can use the leftovers in harvested fields, or crops specially provided for their use (lure crops) (378, 379). Gas cannons work well on species that are hunted, because they associate the noise with danger. Gulls are not hunted and do not normally react to gas cannons.

Although resident birds at airfields soon become accustomed to unnatural sounds, it seems worthwhile to use gas cannons during the migration seasons, when large numbers of birds unaccustomed to aircraft may come in late at night and cause serious problems when flying starts in the morning. To delay habituation of the birds, the explosions should take place at an irregular rate and the position of the gas exploders be changed frequently. In Germany, a slowly rotating gas cannon is used, so that bangs are given in all directions, which is supposed to increase its effectiveness.

The other two categories of sound, natural and synthetic, are more effective and are discussed in greater detail.

(c) Natural sounds

Bio-acoustics is the science concerned with animal sounds, and the mechanisms to produce and hear them. This section deals with the use of bird sounds to influence their behaviour.

Like humans, birds use certain postures and movements ("body language") as well as vocalizations to communicate. The vocal utterances of birds consist of songs and call-notes. There is no sharp distinction between songs and calls, but call-notes almost never consist of more than four or five notes (405). Bird sounds have been the subject of several books (e.g., 20, 155, 404).

Song is primarily of importance during the reproductive season (establishment of territory and attraction of females), whereas most calls or call-notes are important survival cues throughout the year. The classification and identification of bird calls is far from complete at present, and the terminology is often confusing (156). Bird call-notes have been classified as: pleasure calls, distress calls, territorial-defence calls, flight calls, feeding calls, nest calls, flock calls, aggressive calls, general alarm calls, and specialized alarm calls (404).

The calls of some nuisance species have been studied intensively. The starling, for instance, reportedly has 11 different calls, 3 of which are "danger calls" (195), and the Red-winged Blackbird has 6 "general alarm calls" (the "chuck," "cut," "check," "chick," "growl" and "scream") (319).

Most gregarious birds have (1) alarm calls to indicate the presence of a predator in sight (the calls are innate and are different in each species), (2) alert notes, warning of impending danger at a distance, and (3) distress calls, given when seized by a predator or taken from a net, immediately after capture (62).

Recording these calls is not easy and often requires a lot of patience. Distress calls of gulls can be obtained by taking the birds in the hand, immediately after one has trapped them (Fig. 5-2). Once a gull is resigned to its state of captivity it is hard to provoke any more distress calls. Alarm calls and alert notes can be recorded in experimental set-ups where wild birds are exposed to faked

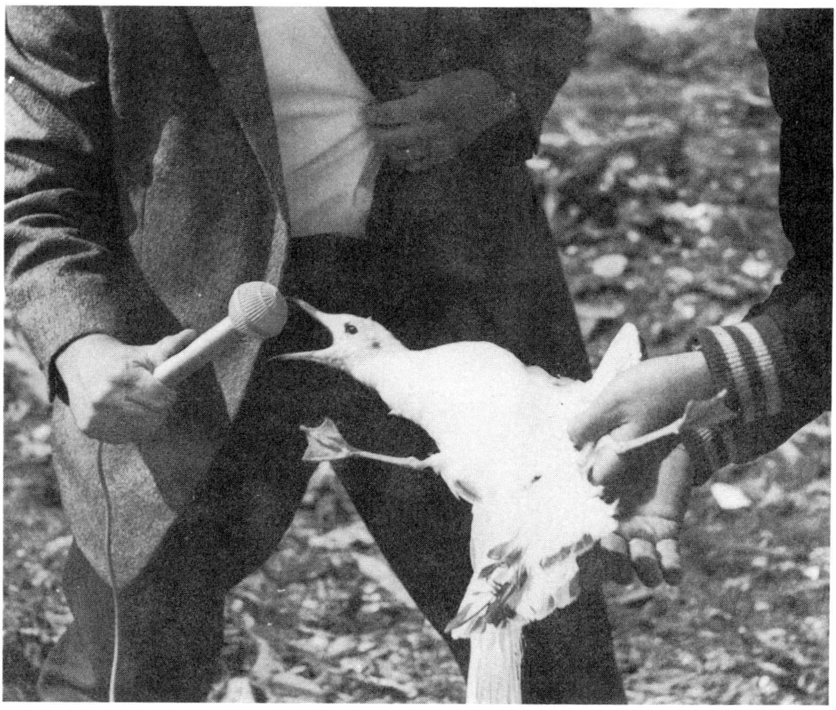

Fig. 5-2: The distress call of a newly captured gull is being recorded for use in driving gulls from airports.

threats (e.g., a stuffed cat near a bird nest). Distress calls can also be obtained by the use of electric shocks (249).

Both distress and alarm calls have been used to frighten birds away. The reactions to broadcast distress calls vary with the species. When their distress call is played, gulls usually become alert, fly towards the loudspeaker, climb, and gradually disperse when the broadcast ends (or shortly before). Corvids (Carrion Crows, Rooks, Jackdaws, and Magpies) behave more or less in the same way, but starlings usually fly directly away on hearing their distress call (70). Reactions to alarm calls also show considerable diversity, but in general, birds leave the area at once rather than first investigate the source of the sound. For this reason, alarm calls appear to be more suitable for use at airports.

Although the call-notes are basically characteristic for each species, some species may learn to respond to calls of other species with which they live in close association (62). This interspecific response often occurs among closely related species, e.g., broadcasts of Herring Gull calls will disperse Black-headed and Common Gulls, and vice versa. It was concluded from work in Britain (70): "It may be that some of these birds respond to the flight behaviour of their neighbours rather than to the call itself, nonetheless the reaction is beneficial for scaring purposes."

Although for many nuisance species the alarm and distress calls have been recorded and tested, some species have no alarm call (such as the Laysan Albatross (62)), and others appear to have no distress call (such as the Wood Pigeon and Oystercatcher (70).

Birds have individual songs and calls that can be identified by other birds, and the calls of one individual may be more effective in scaring than that of another (67). It has been suggested that birds have difficulty in understanding the "dialect" of birds of the same species from other locations, which would explain a failure to disperse starlings in England with the distress call of an American starling. However, in more recent work this problem of dialects was not encountered, so in all likelihood it is of minor significance if it exists at all (70).

There are many factors that influence the response of birds to these species-specific bio-acoustic stimuli: time of year and day, physiological state and activity of the bird, weather conditions, previous experience, and the quality of sound.

During the breeding season, or when a bird is hungry or tired, it may show a less-than-normal response to broadcast calls. Starlings that have safely settled in their roosts are hard to frighten away with distress calls at night, but have been successfully repelled during their arrival at dusk (71).

Birds do not become accustomed to natural calls as easily as to non-natural sounds. Nevertheless, birds will gradually learn that their broadcast distress or alarm call does not constitute a genuine hazard and may stop responding. Birds take longer to become used to good-quality sounds. Changing the rate of playback and the position of the speakers, and alternating several calls also helps, but the best way to prevent habituation is to reinforce the calls with an additional, preferably visual stimulus, like a smoke puff or shellcracker.

Although some authors have reported habituation to distress calls (76, 77, 78, 170), the following phenomenon has also been described: Band-tailed Pigeons harassed by gunfire, shellcrackers, and their protest sounds (sounds not actually of distress or alarm), began to respond gradually, and it was only during the third day that many began to fly away from the field where they were feeding (62). From this and other experiments (including other species), it was found that about four days of exposure to alarm sound is necessary before birds that cause crop damage will respond consistently to their own sounds and will have their numbers greatly reduced (62).

To get the message across to nuisance birds, the taped call and the playback equipment should be of reasonably good quality. During a joint project of French and English researchers, it was found that the results were similar with both high- and low-fidelity equipment under favourable conditions, but under adverse conditions, especially of wind, the high-fidelity equipment gave better results (67). A detailed account of what equipment to use for recording bird calls and for broadcasting them is given in (156).

Bio-acoustic bird dispersal methods have been tested at airfields in several European countries, including the USSR, and in Canada and the US. Holland was one of the first countries to experiment with distress calls at airports. At some military fields two rows of loudspeakers were installed permanently along the main runway, and distress calls of three gull species and the Lapwing were used. Initially, good results were obtained (185), but in later years the effectiveness of the installation decreased to almost nil. This may have been due to habituation and to deterioration of the sound quality as a result of modification of the loudspeakers, which became necessary to prevent rupture of the vibrating membranes in the speakers when noisy Starfighter aircraft were introduced (126). At present the system is not in use.

At Schiphol Airport near Amsterdam, distress calls (in combination first with flares and later shellcrackers) have been used since 1955 with satisfactory results (421, 422). Vans are fitted with a cassette-type tape-player and loudspeaker on the roof. Bird-clearing operations are carried out regularly and are part of the responsibilities of the airfield inspector.

In the UK, the frightening of birds by acoustic means has gradually developed from scattered experiments (28) to a full-sized routine operation (70, 444, 445). At all RAF airfields and at most international airports the firemen have been supplied with distress-call equipment and a manual on how to use it (412), and have been given lectures on bird identification and behaviour. Proper training and instruction is essential for successful operations. The UK acoustic system is complemented by shellcrackers. This combination has proved successful in practice and the birds have not become used to it. A broadcast system could also be used to play back various loud noises (clapping hands, banging metal on metal, etc.) possibly mixed with the bird calls (156). The use of the sound equipment as part of a public address system is an additional advantage. Britain has serious bird problems but is fortunate in that the main pest species readily react to their broadcast distress calls. In other countries with different problem species, the system might be less effective (445).

Development of a bird-scaring system based mainly on the broadcasting of bird calls must include studies to determine (1) what species cause what problems, at what time of the year and of the day, and under what conditions, (2) what calls can be used to scare the birds and how these calls can be obtained, (3) the minimum quality requirements in both the taped call and the playback equipment to obtain satisfactory results, (4) the best rate of broadcasting, (5) the best additional stimulus (visual or acoustical) to reinforce the effect, and (6) what else can be done to prevent habituation. Clearly, the necessary research and development work should be carried out by biologists and may take much time and money. The final results, however, may well be worth the investment.

Although natural bird sounds are numerous, only two (distress and alarm calls) have been used for routine bird-scaring purposes. It may be possible to use other calls as well, or to flood a certain area with so many natural bird sounds that the acoustic communication system of the birds becomes inadequate and the birds prefer to go elsewhere (156).

The calls of hawks and owls, and also bird calls that attract and identify the species (e.g., food finding, mobbing, assembly, and courtship calls) may have value in bird control (156). During tests at Vancouver International Airport the calls of a Peregrine Falcon drove away gulls (178). At this airport Dunlins (Red-backed Sandpipers) winter in large flocks and have resisted removal by conventional techniques. They have caused several serious aircraft incidents and are an extremely persistent problem during fall and winter. Experiments are underway to remove them by disrupting their voice communication through broadcasting some of the calls they utter when feeding (178).

(d) Synthetic sounds

Sounds can be analyzed by modern equipment to show their frequencies and intensities. The "voice-print," the acoustic counterpart of the fingerprint, may help to identify not only the sound of a species, but the individual bird that made the sound (62). Conversely, it is also possible to synthesize sounds.

The possibilities of synthetic sounds are obvious: natural sounds can be distorted and completely new sounds can be created in order to produce a "super stimulus" that may elicit a "super response" in the birds.

In a test with House Sparrows, Boudreau (62) obtained a 94% response to their natural alarm sounds, and 83% to the synthesized versions of these sounds. He also successfully used synthetic alarm sounds on starlings. In another experiment he recorded and later synthesized the compressor noise of T-37 aircraft and broadcast these sounds to Red-winged Blackbirds flying to and from their roost. At levels above 73 dB the blackbirds detoured widely around the sound source. The synthesized sound had a greater effect than the original sound.

COMBINED VISUAL AND ACOUSTICAL SCARING METHODS

(a) Pyrotechnics

Pyrotechnics (fireworks) used at airfields to disperse birds should be effective, safe, and operationally feasible. In frightening birds it is usually extremely difficult to get complete success at all times. If a certain type of fireworks gives satisfactory results most of the time on most species, it should be considered as a useful tool, rather than a less-than-ideal device. Regarding safety, the fireworks should be easy and safe to operate, they should give no confusing information to pilots, should not start fires in grass or buildings, and should not leave any debris on or alongside runways and taxiways that could be sucked into and damage jet engines. The main factors in operational feasibility are cost and manpower.

Pyrotechnics used for bird control at airports have included modified Very flares, shellcrackers of various types, and signal rockets.

A Very pistol (invented by E.W. Very) is normally used for firing coloured signal flares (Very flares) and smoke puffs at airports (Fig. 5-3). The cartridge of a Very flare is much bigger than the shell for a shotgun. For use in bird dispersal the smoke puff is replaced by a firework that will explode a set number of seconds after firing the projectile. When fired at a 45° angle, the explosion should take place at the end of the trajectory, just above the ground. The explosion produces a bang, a flash of light and some smoke, and it disintegrates most of the firework. However, the charred remains of the firework might be a source of foreign object damage (FOD) if they happen to fall on the runway. These Very fireworks were used in combination with distress calls at Schiphol Airport with good success, until they were replaced by specially designed debris-free shellcrackers (421, 422).

Shellcrackers are much smaller and cheaper than the Very flares and can be fired from a shotgun or a pistol (Fig. 5-3). Here too, the projectile contains a firework that explodes at the end of its trajectory. The problem is to pack a great range and a big bang in a safe cartridge. Some mishaps have occurred with shellcrackers due to premature explosion in cases where a choke-bored shotgun was used (most shellcrackers are designed for cylindrical barrels). Shellcrackers should be reliable, stored with care, and used with caution. Shellcrackers now used at Canadian airfields were tested by military and civil explosives experts, who also checked for absence of debris.

In some countries shellcrackers are fired from a Very pistol that has been equipped with an adapter barrel. In West Germany, a blank cartridge pistol is used to fire a small, cheap shellcracker (Fig. 5-3). The sky is the limit when it comes to different designs of fireworks. In the UK the "star hummer" was developed and tested at airfields. This variant on the shellcracker has an erratic flight, a high-pitched scream and, as climax, a loud bang (445). Despite these impressive features, its success in the field is not spectacular, and most UK airfields continue to use the standard shellcrackers in combination with distress calls (75).

Shellcrackers are used at many airports with varying degrees of success; habituation is the main problem. The official instruc-

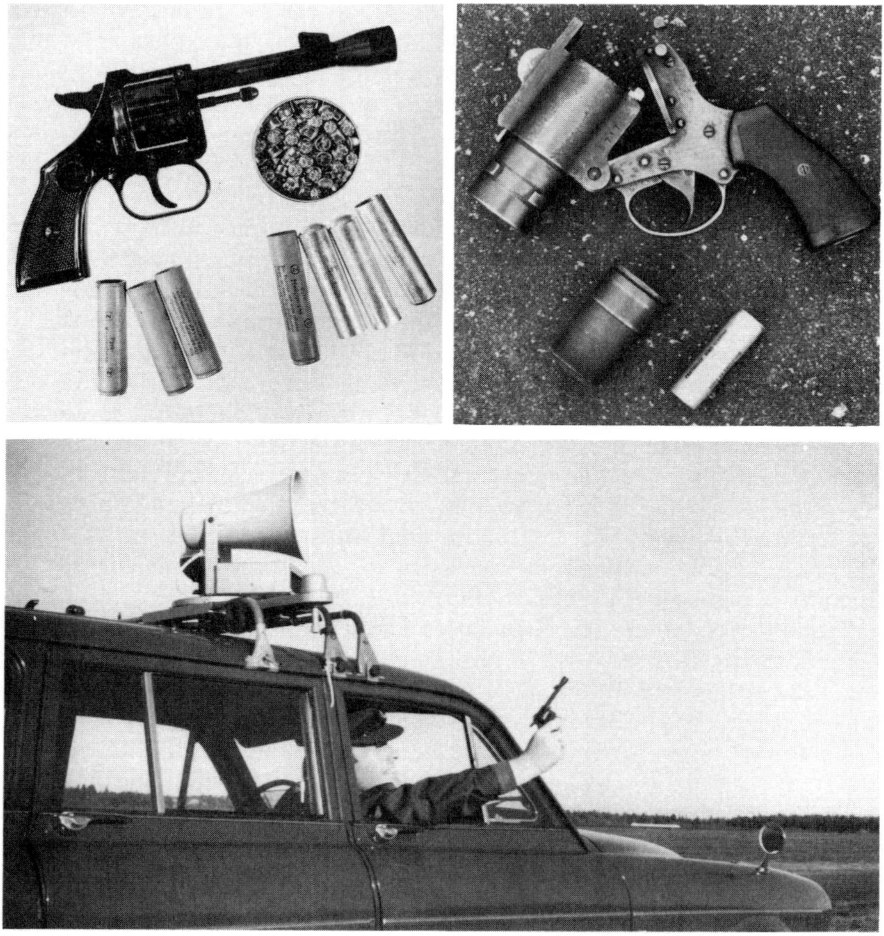

Fig. 5-3: Various methods are used to rid airfields of birds. (top left) A German pistol and small shellcrackers used for frightening birds. (top right) A Very pistol and adapter barrel used for firing standard-sized shellcrackers. (bottom) Distress calls are broadcast in combination with discharge of shellcrackers—an effective means of clearing birds.

tions for bird control at civil airports in Canada recommended that when shellcrackers begin to lose their effectiveness, occasional live shooting of a few birds may be necessary. "However, wholesale slaughter is both ineffective and unwise" (95).

Signal rockets may be bigger than either shellcrackers or Very flares, have a longer range, may require special launching equipment, and are considerably more costly. Tests in the USSR with a special signal rocket showed that it had an effective range of about 1,300 feet; however, most birds soon returned and settled again on the spot from which they had been scared (234). An additional effect was created when after the explosion, a number of coloured flames fell towards the ground, emitting a howling sound. The GAF manual on bird control at airbases mentions the "Spezial-Blitz-Knall-Rakete" (special flash-and-banger rocket), and recommends that different pyrotechnics be used in an alternating manner and from different sites, to avoid habituation of the birds (201).

(b) Birds of prey

The use of raptors to clear birds from airfields has attracted public interest. It is indeed unusual when a medieval sport is used to safeguard jet-age air travellers.

Falconry is the art of taking wild quarry with captive birds of prey. Members of the falcon family (Falconidae) and the true hawk family (Accipitridae) are commonly used. Buteos or buzzards, often called "hawks" in North America, are infrequently used in falconry. Falcons have long tails, long narrow wings and are birds of the open spaces (moors, marshes, prairies, deserts, etc.). Accipiter hawks have short tails, short rounded wings, and are birds of the woodland and scrub (354).

Falconry is a specialized form of hunting, and books have been written on its techniques and terminology (e.g., 49, 441).

Wild falcons quickly climb above their prey and swoop down on it, whereas wild hawks chase their quarry in a rapid pursuit, often between trees and through bushes. The methods that the falconer uses are adapted to the different raptor and prey species.

Experimental work at airports began in 1947-49 in Britain, with Peregrine Falcons. From these trials it was concluded that an airfield can be kept clear of birds by flying a Peregrine Falcon at least once a day, but the clearance does not last for more than two days once the falcon is taken away. Major drawbacks to the use of this method were: keeping and training the birds was time-consuming and costly; some airfields were unsuitable for falconry (birds got lost or were shot); and the falcons could not be flown in darkness or in bad weather such as fog, heavy rain, or high winds (443).

Nevertheless, a falconry program was started at Royal Naval

Air Station Lossiemouth in Scotland, because its serious problem of local breeding gulls could not be solved by other methods. The Peregrine Falcons scared the gulls away during the day and in good weather. Gulls that returned at dusk were harassed by firing shell-crackers at irregular intervals during the hours of darkness. Furthermore, Thunderbird exploders were used at the ends of all runways during aircraft operations. After two months, the gulls had left the airfield and found a roost elsewhere. The operation has been highly successful: no bird strikes have occurred since establishment of the falconry section and after two years there were few, if any, birds on the field (197).

In Canada, large numbers of Glaucous-winged, California, and Mew Gulls frequenting Victoria Airport on Vancouver Island provided ample opportunity to test the operational feasibility of falconry at airfields. A falconer and assistant flew Peregrine Falcons from 1962 to 1964, and Gyrfalcons from 1964 to 1965. The gulls were invariably dispersed from the area but frequently returned soon after the falcon was back in its cage. When winter rains brought many worms onto the tarmac, several falcon flights per day were necessary to offset this attraction.

At another airport, near Halifax, Nova Scotia, a falconer, also using peregrines, tested a different scaring technique: the falcons did not approach or attack the gulls but instead circled high above the falconer until recalled by him. These experiments on Herring and Great Black-backed Gulls, in the fall of 1964, showed that this method was effective, that the falconer had better control over the bird, and that the falcon would not be injured.

Despite these satisfactory results the Canadians decided not to use falcons because of the limitation to daylight hours (many strikes occurred at night); the need for trained and dedicated operators with a radio-equipped vehicle; the requirement of a dependable supply of falcons; and the availability of other, and less costly, scaring methods for use by untrained staff (365).

The Dutch military tested the effectiveness of Goshawks to control a serious gull problem at Leeuwarden Airbase. A falconry team (consisting of 1 falconer, 3 assistants, 6 hawks, and a radio-equipped jeep) worked on the airfield during most of 1968 (Fig. 5-4). Results were satisfactory (fewer bird strikes compared to previous years), but after a while the gulls tended to fly a short distance away on approach of the falconer's jeep before the hawk could be used. In such cases smoke puffs and shellcrackers were used (362). Although the hawking group proved its limited effectiveness, it is impossible to say how much was due to the

Fig. 5-4: Birds of prey, such as these Goshawks, have been trained successfully to frighten birds from airports.

presence of the patrol group and how much to the hawks (342). The hawking group was disbanded because of the high cost of staffing a permanent falconry unit.

In Spain, six Peregrine Falcons were used to get rid of thousands of Little Bustards that caused serious strikes at Torrejon Airbase near Madrid, a busy airbase used by USAF Europe. After three months of falconry operations all nuisance birds had left the airbase and no more strikes occurred (113); but the falcons still had to be flown every day to prevent the birds from returning (339).

Because of these good results, peregrines were introduced at the Barajas-Madrid civil airport to drive out Little Bustards, Stone Curlews, and Mallards. The success of clearing bustards from Torrejon had increased the bird problem at Barajas, less than five miles away (366). Again, the falcons proved to be effective: after six months both runways and the airfield were completely free from these three species (339).

Encouraged by the results in Spain, USAF Europe decided to employ falcons at its six bases in the UK, where a full-time year-round program was introduced in 1970. At the airfields mainly Saker, Laggar, and Lanner Falcons are used, all of which are imported. (Peregrine Falcons can no longer be obtained in Europe.) Other scaring techniques are also used, including dogs, shell-crackers and live ammunition. Although 80% of the bird-clearing work is done without falcons, the 20% with falcons is considered essential (118).

The USAF has begun a falconry program at an airbase in Turkey and also tested falcons at Whiteman AFB in Missouri, US (286).

When reviewing the various past and present falconry activities at airfields, a few facts become clearly evident: (1) properly trained birds of prey of the right species for the job at hand, used regularly and persistently by skilled and conscientious personnel, are effective in clearing nuisance birds from airfields during daylight and good weather; (2) for good results, daily operations on a year-round basis are required in most cases; (3) several falcons are required in order to have at least one falcon on standby at all times, ready for action; (4) to obtain, train, operate, and care for falcons a staff of at least two full-time well-trained personnel is required.

At present in most western countries a permit is required for the capture of wild raptors, and some species may not be captured at all. The use of raptor species that are threatened with extinction is of great concern to environmentalists (304, 315) and it should not be encouraged as the standard method for all airfields with bird problems. The world population of these raptors would be unable to sustain the required supply without causing further decline of their numbers.

Recently this situation has changed a bit. When raptor numbers began to decline disastrously, attempts were begun to breed birds of prey in captivity. Based on a new technique that can be described as "co-operative, artificial insemination," propagation of captive birds has been successful for several species including the American Kestrel (326), Prairie Falcon (86), Peregrine Falcon (86), Goshawk (38), Red-tailed Hawk (395), and Golden Eagle (171).

So far, most of the successful experiments to breed raptors in captivity have been carried out in North America, particularly at Cornell University. If this know-how improves and if similar results are obtained in Europe and elsewhere, there might well be a limited but sustained supply of falcons for use at airports. At present, Canadians are considering use of falcons bred and reared at Cornell, for experiments at Vancouver Airport against huge flocks of wintering Dunlins (266).

(c) Radio-controlled model aircraft

As the use of real raptors has some shortcomings, an engineer may feel challenged to build something better. An attempt to do so was made in New Zealand, where a remote-control model aircraft was used against the nuisance birds at Auckland Airport. The model had a wingspan of 5¾ feet and a length of 3½ feet and was painted so as to resemble a bird of prey. The standard engine was radio-controlled within a quarter-mile radius of the operator. Preliminary results were promising (345).

In Canada, preliminary work was done with a model airplane to move Dunlins from the salt flats near Vancouver Airport. The tests were successful in causing the birds to move, and further experiments with a falcon-shaped model are in preparation (Fig. 5-5).

The Canadian Wildlife Service gained experience in the use of model aircraft when attempting to scare robins from low-bush blueberry fields in New Brunswick. Robins could be flushed and driven off by the noisy aircraft. But other species, such as sparrows, waxwings, and swallows, did not appear to be bothered. The robins came back to the blueberries as soon as the airplane had

Fig. 5-5: Radio-controlled model aircraft built to resemble a falcon in flight are being tested in Canada for their effectiveness in clearing birds from airports. The model shown has a flying time of over 30 minutes and can be manoeuvred realistically at ranges of up to half a mile.

landed, making continuous operations almost a necessity. Another shortcoming was the need for skilled operators to control the model (158). Where skilled operators are available, remote-control aircraft may prove useful in dealing with a persistent bird problem.

CONCLUSIONS

A mobile, aggressive, imaginative, and persistent patrol group will be able to get rid of most birds at most airfields. A patrol team responsible for keeping the field free of birds needs a radio-equipped vehicle, shotguns, shellcrackers, searchlights, distress call equipment, gas cannons and, above all, strong motivation. Only if such a team should fail to obtain satisfactory results, even after a prolonged period of persistent harassment of birds, would it be justified in embarking on more costly programs such as falconry and remote-control aircraft that require specialists and that are limited to daylight hours and reasonably good weather.

BIRD REMOVAL AND BIRD KILLING METHODS

If nuisance birds at an airport cannot be scared away by any of the methods mentioned above, two options remain: capture and removal from the airport, or killing. The first possibility requires a great deal of work and the released birds may return to the airfield. Nevertheless, this method is often justified, most clearly so in cases where rare or endangered species are involved. Practical ways to kill birds are trapping and dispatching, shooting, and poisoning. In many countries a large number of bird species are protected, and killing, trapping or transporting wild birds is subject to regulations. However, permits can usually be obtained for specific purposes.

Bird trapping has been practised through the ages to obtain food, to get rid of objectionable birds, and to capture desirable birds. The variety of traps and trapping techniques is wide, ranging from small, baited spring-nets (operating much the same as mousetraps) to huge contraptions complete with a battery of light bulbs to attract the birds and a chamber to gas them (142, 296). A thorough account of bird trapping and banding is given in a series of four books in German (83). References in English are found in standard textbooks (325, 389). Special techniques have been described for live trapping of hawks and owls (290, 337).

Shooting of birds on airfields is not readily accepted by airfield administrators. In addition, the use of firearms against birds requires special permits.

When poisons are used, there are at least four categories of substances to consider: (1) poisons that can be eaten or inhaled, or that can penetrate into the body directly through the skin; (2) narcotics which make the birds drowsy and sleepy in low doses, but kill them in high doses; (3) "flock alarming" chemicals that produce serious distress in birds, and induce flock members to leave the scene; and (4) chemicals to sterilize eggs.

For many years the US has been developing chemicals to control bird damage to agricultural crops and to solve non-agricultural problems in urban and suburban areas (346). Papers reporting some of the results of these investigations are available from the Denver Wildlife Research Center on request. There does not appear to be an up-to-date comprehensive manual on the practice of chemical bird control.

The use of chemicals generally should be discouraged because of the problem of getting the right dose into the right bird at the right time and place. Poisoning operations may affect or kill non-target species. Any poisoning operation should be supervised, and preferably carried out, by a licenced expert.

The elimination of a local bird problem by eradicating all birds present will be at best only a stop-gap measure, because other birds in the airport's vicinity will soon move in to fill the vacant area. A more permanent solution can be obtained only by finding out what attracts the birds to the airfield and by removing or destroying that attraction. This approach is discussed later in this chapter.

BIRD CAPTURE AND REMOVAL FROM THE AIRPORT

At two Canadian airports birds have been captured and released elsewhere on a routine basis.

At Vancouver Airport, owls and hawks were a serious problem as they fed on the numerous voles that lived in the areas with rough ground and high, unmown grass. Several of the 11 predator species recorded at the airport had been involved in collisions. All attempts to scare the birds had failed. It was decided to live-trap and remove the birds of prey, because shooting would be time-consuming and would destroy useful birds. During a period of almost three years, Verbail pole traps were used to live-trap the birds of prey that frequented the airfield. This device is installed on top of a pole and employs a snare to hold the feet of any bird that is heavy enough to trigger the trap. (Construction details are given in 224.) A total of 543 hawks and owls were caught, banded, and released at distances of up to 40 miles from the airport. Of the 426 Short-eared Owls released, 23 were retrapped later on the airport, and three were retrapped once more. Eight of the banded owls were later found dead on the airfield, believed to be victims of bird strikes. Of all 426 banded owls removed from the airport, 7.3 per cent were later caught or found dead on the airport. Although this percentage seemed quite high, it was deemed more advisable to live-trap and remove the birds than to destroy them (218).

At Toronto International Airport, wintering Snowy Owls hunt for voles along the runways and taxiways, which are kept free of snow (107). Serious and costly collisions occurred and airport personnel were given permission to shoot them. When it became known that owls and hawks were being shot at the airport, local conservationists volunteered to trap, band, transport, and release the birds. Airport management approved the plan, which proved to be a great success and which is now financially supported by the Ministry of Transport.

KILLING OF BIRDS

Trapping and dispatching

Crows have been trapped on a routine basis at Frankfurt International Airport and a few other airports in West Germany. The seven traps installed at Frankfurt caught an average of 1,000 crows annually. Trapped crows were killed by a blow at the base of the skull (breaking the neck), by poison, or by exposing the birds to chloroform in an airtight box (201). Hawks and owls found in the traps were released 30 miles from the airport.

Sparrows and pigeons in hangars can be captured with suitably baited traps located at their normal feeding sites.

Shooting

Shooting with live shells can be used to scare birds, to kill birds, and to reinforce other stimuli such as distress calls. When shooting to kill, ammunition should be heavy enough to kill quickly. The following table provides a guide to shot size (224):

Birds	Size of shot						
	Diameter		Commercial designation				
	(in.)	(mm)	English	American	Belgian	French	German
Geese	.18	4.5	B B	Air rifle	—	—	—
Ducks	.13	3.25	3	4	3	4	4
Plovers	.09	2.25	8	8	8	8	8
Smaller birds	.05	1.25	(———— 12 (or smaller) ————)				

— unknown or not given

Guns must be used judiciously, and proposals to patrol airfields by gun-equipped personnel should be carefully considered by the airport manager's office.

It is often difficult to get rid of birds in hangars. The use of shotguns around airport buildings is usually prohibited, but at an airbase in the UK .22 air rifles were used to do away with nuisance birds in a hangar (385).

Killing and removal with chemicals

The use of poisons at airports has been most successful in dealing with "urban birds" in hangars. Where live trapping is not feasible, a baiting area where food is laid out for the birds should be developed. Once the birds have formed the habit of feeding on the bait, part of it is treated with a quick-acting poison. Dead and dying birds should be collected and disposed of to prevent scavengers from feeding on the dead birds and becoming affected as well. Selection of an attractive bait and proper use of the right poison usually requires experience and judgment, and this work can best be carried out by professional pest-control experts.

Poison can be given to the birds via their food, but it can also be administered as a "contact poison," the toxicant penetrating directly through the skin. Birds have little bare skin, but when they alight on a treated area they receive the poison via their feet. The poisonous chemical can be applied directly on all areas on which birds land, or it can be contained in artificial perches that are installed in the entry ways to hangars and near other areas that the birds frequent. The perches may have an interior reservoir and a wick that gradually feeds the chemical solution (endrin) from the reservoir to the surface of the perch. As contact poisons are highly hazardous to man as well, it is good safety procedure to remove the perches when bird numbers have been reduced to an acceptable level. The chemical endrin has also been used to spray entire hangars at two US airbases. The method worked well (birds were killed), but as the material remains toxic for two years, the USAF has stopped its use because of the danger to personnel (363).

Another contact poison that has been used successfully for bird control in hangars is called Queletox. Adhesive tape is fastened to the edges of access holes to the hangars, and the toxic material is applied to the tape with a caulking gun. This technique was used with good success at four airbases in the US. Most birds (sparrows, starlings, and pigeons) died inside the buildings, but some were found outside a few hundred yards away. The tape and toxicant were removed after the operation. Using an electric lift, the crew took one afternoon to treat one building. One application was found to be effective for three or four months (363).

A narcotic has been used in a highly successful operation to destroy Southern Black-backed Gulls breeding in three colonies in the immediate vicinity of Napier Airport, New Zealand. The chemical alpha-chloralose was used in high doses on bait so that birds would die relatively quickly in their sleep and with little disturbance. Other poisons were considered more painful (causing distressing convulsions) or less effective (the birds regurgitating the bait). After four days of prebaiting with unpoisoned bread distributed by aircraft over the colonies, narcotic-treated baits were used. Of the estimated 2,500 breeding gulls present, more than 85 per cent were killed, and the operation was followed by a spectacular drop in gull strike numbers (87).

If a "flock-alarming" chemical is eaten by a few birds in a flock, their abnormal behaviour disturbs others in the flock, causing them to fly away. Avitrol 200 is the commercial name for a chemical (4 amino-pyridine) that causes birds to act "strangely." Excellent results have been obtained with Avitrol 200 (applied to whole corn) on an estimated 10,000 crows roosting on Friendship Airport near Baltimore, Maryland, US. All birds left and none had returned six months later. It was also used successfully to control loafing gulls at a naval air station at Norfolk, Virginia, US (166), and to clear hundreds of pigeons from a hangar at Montreal International Airport.

Egg sterilization by chemicals has a place in bird control. Gull control is not easy, and a recent review of gull management methods concluded that there is no record of an effective method for reducing gull populations over large areas. There are basically two methods to reduce a gull population through control of their eggs and young: removal of eggs, and sterilization of eggs. The latter method is often preferred for large areas containing numerous nests. If eggs are removed or broken, gulls will lay again, making it necessary to repeat the procedure until the urge to nest has passed. When eggs are sterilized by pricking, shaking, or chemical treatment, the trick is to prevent both hatching out and re-laying. Eggs in which the embryo has been destroyed will rot and burst, causing the birds to re-lay. Thus it is necessary to add a preservative to the sterilizing chemical, so that the parent birds will continue to incubate the lifeless eggs for a sufficiently long period of time beyond the normal 3½ to 4 weeks that they will be physiologically incapable of laying a new clutch that year (400).

At Copenhagen International Airport, Herring Gulls are the main problem birds. Since nearby Saltholm Island supports a breeding population of over 20,000 pairs of Herring Gulls, killing a few hundred gulls at the airport would obviously be no solution because others would continue to be attracted to the airfield, which they use for loafing and feeding. Thus it was decided to eliminate the gulls at Saltholm by spraying their eggs. Since 1969 an annual spraying campaign has been carried out (with an emulsion of oil in water containing 10% formaldehyde) at Saltholm Island. The operations covered all of the 6 square miles of the island and involved 15-18 people working for 10 days. Herring Gull nests have decreased from 34,000 in 1969 to 21,000 in 1973 (272). As Herring Gulls have a long life (15-20 years, occasionally over 25 years), the annual operation will have to be continued for several more years, until all members of the colony have died. The problem is, of course, that although each year fewer gulls of the original gullery remain to nest, the island becomes more attractive to surplus birds from other breeding areas. To what extent immigration will offset the hard-gained results remains to be seen.

The Herring Gull has also become a nuisance bird along the Atlantic seaboard in northeastern North America. In 1876 the Audubon Society saved this gull from being exterminated, and in 1969 it discussed a strategy that might be used to manage the 135,000 pairs nesting on 270 islands in the eastern United States (136). Of the control methods considered, the most promising appeared to be the spraying of eggs for several years. However, if this method was applied alone, it would tend to redistribute gulls over previously unused breeding areas, because birds that fail to hatch eggs for several years in succession may shift to new areas. Hence the spraying program would have to be combined with selective elimination of those birds that shift to new colonies. At the same time, similar programs would have to be carried out at breeding colonies along the east coast of Canada in order to deal with the whole Atlantic population of North America.

Conclusions

In theory, capture and removal of birds is time-consuming, costly, and largely fruitless because the displaced birds will return, or others will move in to fill the vacant area. In practice, volunteers with a keen interest in birds and working with dedication and persistence have achieved useful results. This method will work for small numbers of "interesting," protected birds (such as raptors), but not for large flocks of common pests (such as starlings).

Large-scale long-term killing operations are likely to be useful in only a few cases. Extermination of a certain colony of locally breeding birds may have greater justification and more effect. However, in general, elimination of birds from an airport by killing will draw in new birds with less "aircraft experience," which are thus potentially more hazardous than the birds they replace. Occasional killing of a few local pest birds that are not protected may be necessary to prevent habituation against standard bird dispersal techniques.

Methods to control bird problems more permanently, through altering the airport environment, are discussed in the following section.

HABITAT MANIPULATION

As long as an airport retains the features that make it attractive to birds, scaring and killing will remain necessary, because the birds will always come back or other birds will move in. When the attractions are removed, bird numbers will decrease permanently and bird strikes will occur less frequently. That is the reasoning behind various programs to make airports and their vicinity less attractive to birds. This can be done by changes in the physical environment of the airport, or, in the terminology of wildlife managers, by habitat manipulation. Birds are present at airports because airports have bird habitat. By manipulating the habitat it is possible to alter numbers and species composition of the birds.

The main attractions that airports have for birds are food, water, shelter, safety, and places to nest, rest, and roost. The trick is to remove these attractions as much as possible without introducing new attractions for other species. Many bird species are rarely involved in bird strikes, whereas others are a serious pest (described in Chapter 2). A robin nesting beside the hangar is of no concern, but something should be done about a flock of gulls loafing at the head of the runway. Thus, habitat manipulation should be aimed primarily against those species that are the greatest hazard because of their size, flocking habits, or behaviour with respect to aircraft.

ECOLOGICAL RESEARCH AS BASIS FOR ACTION

Ideally, an ecological study covering all seasons should be made before any major habitat changes are carried out. The ecological investigation of an airport and immediate vicinity should indicate how many birds are in the area, which species are involved, how the birds are distributed, and why they are there. Observations should be made during all hours of the day and during all seasons to take into account daily and seasonal fluctuations. Studies should include the geography, hydrology, soil, climate, vegetation, bird and other fauna of the area, as well as human activities such as agricultural and waste disposal operations. The research should provide the factual information needed to understand why nuisance birds are present at the airport and should suggest what can be done, in terms of habitat manipulation, to improve the situation.

In practice, few airports have been studied by a biologist on a year-round basis, but brief ecological surveys have been made at many airfields all over the world. Occasionally one quick survey is all that is needed to locate an important source of bird problems. Sometimes the major attraction for birds is so obvious (e.g., an open garbage dump near the end of the runway attracting scores of gulls) that no biological expertise is required to recognize it.

Manuals for ecological surveys of airports have been issued in the US (364) and in West Germany (201).

More elaborate studies of the ecology of an airport are concerned with the distribution of the birds over the field and the factors that affect this distribution. Availability of food is one of these factors, and some studies have investigated the food and food preferences of hazardous birds frequenting airports. At Toronto Airport the relation between raptors and their prey is being investigated. In northern Norway the food of gulls feeding on Sola Airbase was studied (23), and in Australia research was carried out on the diet of birds at Mackay Airport (427).

Results have been reported for other ecological work done in Canada (Toronto, 175; Winnipeg, 189, 191; Vancouver, 182, 265, 380), in the US (Boston, 145; Washington, D.C., 356), in Europe (Schiphol Airport, 209, 383; Nice, 119), in Asia (Hong Kong, 340), in New Zealand (Auckland, 89), and Australia (Sydney, 425, 426). These investigations have provided a more precise idea of the attractiveness of airports for birds, and some of these attractions will be reviewed in the next sections.

ATTRACTIVENESS TO BIRDS OF AIRPORTS AND SURROUNDINGS

Both the airport (i.e., the area within the airport boundary) and its vicinity offer many attractive features to birds. Airport authorities have control over what goes on inside the perimeter fence, but have little or no say over activities outside the fence. The birds, however, can and do move freely between airport and adjacent areas. Thus, biologically speaking, we should look at the airport *plus* surroundings, but from an operational point of view we should concentrate on the problems at the airport, because that is where many strikes happen and where changes in the habitat can be most easily carried out.

The airport

RUNWAYS AND TAXIWAYS. The areas with hard surface (aprons, taxiways, and runways) offer some attractions to certain species. Gulls and shorebirds like to loaf on the tarmac, and after heavy rain they often feed on the many worms that have been driven out of their burrows and have crawled onto the runway. Crows have been reported to drop rats on runways in order to kill them (229).

GRASSLAND. Grass is a problem because whether kept long or short, it is attractive to certain bird species.

CROPLAND. On many airports, parts of the field are used for agriculture. Although all crops are attractive to some extent, some crops are particularly favoured by birds.

WASTELAND AND WOODLOTS. Cleared areas that are not cultivated, or otherwise looked after, become covered with tall weedy plants and all sorts of shrubbery, which will support a variety of birds.

OPEN WATER. Any body of open water, be it a puddle on the field after a rainstorm or a permanent marsh in a corner of the airport, is a strong attraction for gulls, waterfowl, shorebirds, and marsh birds.

BUILT-UP AREAS. Big buildings, such as hangars, often provide many facilities for birds to nest or rest. Gulls and pigeons like to sit on roofs and ledges, and sparrows, starlings, swallows, and pigeons have been found nesting in or on hangars.

LANDSCAPED AREAS. At many airports, both civil and military, roads, hangars, and other buildings are made to look more attractive by landscaping. Many ornamental shrubs and trees attract birds, but some do so to a much greater extent than others.

MISCELLANEOUS BIRD ATTRACTIONS. Apart from the various attractions associated with different types of terrain, airports offer some inviting features that are not tied to terrain type. These are garbage dumps, sites to perch and to nest, and remoteness.

Areas outside the airport boundary

The immediate surroundings of an airport may offer various attractions to birds. Particularly troublesome problems arise when birds make regular flights across the airport, for instance when they fly between roosts and feeding areas; and sometimes birds interrupt such flights to loaf or preen on the airfield. The worst problem at many airports is the nearby presence of one or more large refuse dumps that provide free food for a large crowd of regulars, which may then use the airfield to loaf and preen.

After this brief outline of the amenities that are available to birds at and near airports, the next section reviews steps that have been taken at various airports in order to reduce their appeal to birds.

METHODS TO REDUCE THE ATTRACTIVENESS OF AIRPORTS FOR BIRDS

No two airports are ecologically and operationally the same, and no two airports have the same problem with birds. Ecological methods to reduce local bird problems through manipulation of the local habitat vary widely for different airports around the world, and the same method used at similar airports may produce different results. Consequently, it is impossible to give a prescription for a particular problem without thorough investigation. At the same time, a few general guidelines apply anywhere. For example, the airport should be made as much as possible into a monoculture (a vegetation consisting of plants of only one species), thus supporting only a few bird species against which effective scaring techniques can be developed. The greater the variety of terrain the greater the variety of bird species, and the more complex the problem, the smaller is the chance of doing something about it.

The airport

RUNWAYS AND TAXIWAYS. Little can be done in terms of habitat manipulation to make a runway less attractive for loafing gulls. There have been attempts to get rid of gulls by painting their favoured patch of tarmac blue, but this proved unsuccessful. As mentioned earlier, another unsuccessful approach was to lure gulls away from active runways by providing an artificial hot pad near the perimeter of the airport (445).

At Heathrow Airport in England, several gull strikes occurred on a single day after extensive rains had brought the worms to the surface. Sometimes the number of worms on the tarmac becomes so large that the braking efficiency of landing aircraft is impaired (264). As these outbreaks of worms are largely predictable, airport authorities should be prepared to launch a massive clean-up operation with sweepers as soon as the worms begin to invade the runway. A more permanent solution is to prevent worms from crawling onto the tarmac, and three approaches have been proposed: (1) kill the worms by treating the grass strips along the runway with a worm-killing chemical (a vermicide); (2) apply a worm-repellant along the edge of the runway; and (3) install a worm-proof gutter between the grass and the pavement.

At a Canadian airbase, Sevin (carbamyl) and lead arsenate were tested for their worm-killing properties. Sevin proved ineffective, but lead arsenate gave the required results without causing undesirable side effects (contamination of drinking water and "die-off" of songbirds) (440). Although good results were obtained, lead arsenate is a potent and persistent poison and its use is not recommended by the Canadian Committee on Bird Hazards to

Aircraft. A fungicide that kills worms is being tested at Windsor Airport, Ontario, as a more acceptable alternative.

The Canadian Committee has sponsored a search for a worm-repelling chemical to be applied along runways. After several years of research it was concluded that a tar and resin mixture used as a substratum under a layer of poor soil would repel the worms for years, allow the grass to grow, and have no side effects (283). However, because of the high cost involved, the method has not been implemented. The German Air Force recommends the application of a thin band of kainite along the edges of the runway; this is a loose material and can be poured. Effects are likely to be temporary because the "chemical dike" is eroded by rain and wind. The German Air Force also suggested construction of a worm-proof gutter along new runways (201). A worm-proof gutter was tested in a Canadian laboratory several years ago. It prevented worms from crossing, but was judged unsuitable for airfield use because of difficulty in keeping it clean and well drained (374).

Insects on the grassy strips along the runways can be killed by spraying, but if only a relatively narrow strip is treated, insects will soon invade the sprayed area. Because of publicity regarding the undesirable side effects of many pesticides, it need not be stressed here that insecticides should be used cautiously and that their effect both on targets and non-targets should be monitored to determine the usefulness of spraying campaigns.

One reported success was at a USAF airbase where starlings had caused extensive damage to a C-130 on lift-off. The 250 dead starlings found on the runway were part of a large number feeding on crane fly larvae in areas near the runway. These areas were sprayed with a mixture of insecticide (diazinon) and para-dichlorobenzene ("moth crystals"). The latter chemical was added as a starling repellent. Results were good, with the starlings being reluctant to alight in the sprayed areas where their main food source had been killed. The additive may have acted as a bad-tasting or foul-smelling repellent, as discussed earlier in this chapter, but this aspect was not reported (14).

In Australia, it was found that some insect-eating bird species, which had been involved in bird strikes, were foraging at night and in the early morning around runway lights. These lights attracted large numbers of insects, and it was suggested that the white lenses be replaced by orange lights, because orange lights attracted significantly fewer insects. As white lenses were used to conform to ICAO standards, it was suggested that the rationale for having white runway lights be re-examined and the potential of orange lights for reducing the bird strike hazard be further explored (428).

Another attraction that brings birds near or onto the tarmac is the presence of posts, lights, and various markers, spaced along runways and taxiways. Many birds, especially birds of prey, like to sit on a perch rather than on the ground, and most airfields offer few if any other places to perch on. This perching can be prevented by putting sharp spikes on top of these structures or by spraying them with sticky materials that birds dislike to sit on. These sticky materials, exposed to sun, rain and dust, will eventually lose their repellency, and must be replaced.

Apart from these specific measures, it is important to keep runways and taxiways clean. They should be inspected routinely, and all materials that might attract birds (such as dead animals, spilled crops, paper bags, etc.) should be removed as soon as possible.

A bird hazard problem that received much public attention occurred at Midway Atoll in the north-central Pacific Ocean. The atoll consists of two islands (Sand and Eastern), three islets, and some shifting sandspits. In 1956-58, perhaps a third of the world population of Laysan Albatrosses and about a sixth of the world's Black-footed Albatrosses were breeding on Midway Atoll (329). The US Navy constructed an air station on Sand Island shortly before World War II. When air traffic and airspeeds increased, collisions with the seven-foot birds became a serious problem and biologists were called in for advice. Wiping out the breeding colonies would have serious effects on the survival of the Laysan Albatross, a species that had already been destroyed by feather hunters at several of its other breeding colonies.

Driving the tame albatrosses off their ancestral breeding grounds was not effective. Smoke, fire, mortars, bazookas, high frequency sounds, flapping bed sheets, were all unsuccessful: the birds either did not move or soon returned to their nests (8, 245, 328).

The albatrosses were observed to concentrate over sections of runways bordering uneven terrain (dunes, old revetments, small trees), whereas level areas had relatively few birds overhead. Consequently, all land adjacent to the runways was levelled and paved (albatrosses do not nest on hard top), and birds formerly nesting in these areas were killed.

More than 30,000 albatrosses were destroyed, but the control program was not an unqualified success. Its effectiveness remained to be demonstrated, even after nine years. Furthermore, some of the extensive changes in terrain made on Sand Island may have increased rather than decreased the number of birds flying across the runways (148). At present, the problem at Midway is less important because the US Navy has reduced its operations there.

GRASSLAND. The use of grass at airports serves several purposes. Strips along the runway should be firm enough to support an aircraft that gets off the runway and to provide a braking surface. A turf of short-cut grass meets these requirements. These strips along runways should be easy to maintain in order to minimize obstructions and restriction of flying. Grass on other parts of the infield may be grown as a commercial crop (hay, grass pellets, etc.). Grass also holds the soil, prevents dust and dirt from blowing about, requires relatively little maintenance, and makes a pleasing ground cover. Mixtures of grasses with different soil and moisture requirements and winter hardiness are used at different airports.

Long grass harbours ground-nesting birds (partridges, pheasants, ducks, owls, harriers), numerous small mammals (mice and voles), and large numbers of insects. Birds such as gulls and plovers generally do not frequent long grass because it obstructs their view and interferes with movement.

Short grass does not attract ground nesters, nor does it hold large numbers of small mammals and insects. However, it is used by gulls and plovers for feeding and resting, and by small insect-eating birds for feeding.

Observations at Montreal International Airport in 1964 showed far fewer birds on grass of 6 inches or more than on short grass, but at Toronto Airport grass of 6 to 8 inches was consistently chosen by birds for feeding, whereas newly cut grass (4 inches) was often shunned (175). Experimental use of long grass at 10 RAF airfields in the UK, from 1967 to 1969, showed that all birds involved (gulls, lapwings, pigeons, starlings, and corvids) were at least twice as numerous on short (about 2 inches) as on long (6 to 10 inches) grass (74). Later observations have confirmed the preference of the main British pest species for short grass (208).

The decision to keep grass short or long will thus depend mainly on what bird species are an actual or potential hazard. In most areas owls and harriers are not common, whereas gulls and waders are fairly numerous. Consequently, in many countries a long-grass policy is recommended or implemented. In Europe, the long-grass policy is followed at airbases in Denmark (236), Holland (124), the UK (208), and West Germany (201).

In Canada, the recommended grass length is 5 to 8 inches,

both at civil and military airfields (95, 103). Air Training Command of the USAF suggests optimum mowing heights of 4 to 6 inches depending on the type of grass (413). BASH (Bird Aircraft Strike Hazard, the Environics Group of the USAF) recommends grass heights of 8 to 12 inches for several airbases (e.g., 64, 115, 309).

Implementation of the long-grass policy has run into several snags, such as lack of mowing machines that can cut a long sward, financial losses resulting from smaller grass harvest, a possible gradual deterioration of the turf resulting in bare patches that would again attract birds, seeding of the grass, poorer drainage, and the possibility of greater fire hazards. Another difficulty is the need to keep grass short around the runway lights so as not to obscure them; slightly raising the lights might solve this (89). Further, many birds tend to gather on the short-mown strips along the runway when the rest of the infield is covered with longer grass. When this happens, the birds must be scared away.

Although long grass is generally considered the right approach in Europe, at some airports short grass is more likely to improve a problem situation. At Winnipeg Airport on the Canadian prairies, ducks and hawks nested in the grass in 1963 and reared many broods. The young birds were banded and one of the banded birds was later involved in a strike at the airport. In spring and summer 1964, the grass was kept short, and burned in certain areas, in a successful attempt to get rid of ground-nesters (189, 191).

A study of birds and their food at Mackay Airport, Australia, led to the recommendation of short grass along the runways, for the following reasons: if the grass near the runways was left uncut, it would exclude plovers, pigeons, and many other kinds of birds. But hawks and other birds would prey on the grasshoppers and other animals that would wander from the tall grass onto the runways. Grass that is kept short by frequent mowing provides little seed and few insects for birds to feed on.

Once it is decided what grass length will lead to the least bird problems, it may be possible to use a grass mixture that is better suited to being kept at the selected height.

After ecological investigations at Auckland Airport, New Zealand, it was concluded that the ideal solution would be to replace the existing grass cover with a single species of grass that could be maintained by mowing at an even nine inches throughout the year; that would have dense and interlacing growth, so that birds or mammals seeking shelter could not penetrate it; that did not produce seeds; that would withstand drought and remain moist and green, for appearance and so as not be a fire hazard; and that could be controlled in areas where it was not wanted (345).

The Swedish Board of Civil Aviation is testing grass mixtures for the Malmoe-Sturup Airport (411), and the GAF has issued a list of grass mixtures selected on three criteria — little attraction to birds, tough turf, and little annual growth thus reducing the need for mowing (202).

Areas grazed by sheep and cattle are particularly attractive to birds, because the grazing animals disturb many insects, and the animal droppings attract insects. A German study on the effect of sheep grazing on turf pointed out other drawbacks of this practice: the soil becomes harder, less porous, thus making the turf more swampy after heavy rain; sheep droppings produce an uneven fertilization pattern, which makes it hard to keep the entire sward at equal height and encourages the establishment of weeds; and sheep feed only on certain grassland species, making it necessary to mow the remainder.

It is obviously not easy to convince birds that the grass is greener on the other side of the fence. It is also clear that changes in grass management should be preceded by careful investigations and weighing of all the pros and cons and consideration of alternative methods (such as a bird patrol).

CROPLAND. At Schiphol Airport in Holland, just under 20 bird species frequent the airfield, with lapwings, gulls, starlings, pigeons, and partridge present in great numbers for the better part of the year (382). Gulls, partridge, and lapwings were the biggest hazards (most often found dead on the runways). The attractiveness of various types of terrain, expressed as percentage of the total distribution of the birds, is shown in the table opposite that presents the results of a two-year study of the airport. The figures in brackets show distribution during the second year, when the grass was not mown short, but was kept at a height of at least four inches.

In general, croplands were less attractive to the birds than grassland. Although grassland covered only a quarter of the airport grounds, it accounted for almost half of all the birds seen on the airfield. This great attractiveness was slightly reduced (from 48% to 44% of all birds counted) by implementing the long-grass policy. Potatoes, sugarbeets, wheat, and barley were the main crops grown on the arable land (383). Sugarbeets and potatoes

Terrain type	% distribution of terrain type	% distribution of birds All species	Gulls	Lapwings	Starlings
Grassland	25	48 (44)	59 (45)	47 (21)	57 (73)
Arable land	56	42 (54)	18 (52)	46 (79)	35 (27)
Platform and runways	18	5 (1)	16 (3)	2 (0)	0 (0)
Bare ground, rough, ditches	1	5 (1)	7 (0)	5 (0)	8 (0)

attracted few birds. Grains, however, were less advisable because the stubble, which is often loosely cultivated, attracted many gulls and lapwings. All crops were attractive at some stage, especially during harvest and cultivation, and for a short period thereafter. During the months July to October, gulls were attracted equally by grassland and cropland, whereas lapwings definitely preferred the latter.

The distribution of birds found dead on the runway showed a clear peak from July through November (76% of the annual total occurred in that 5-month period). In all likelihood the high attractiveness of the arable land during harvest and soil cultivation contributed largely to this peak. The biologist in charge of the study at Schiphol cautioned that even when types of crops are carefully chosen and agricultural practices are restricted, the fertile soil would continue to attract birds. He concluded that the solution — if any — of the bird problem would lie in an intelligent combination of ecological changes and effective scaring techniques (383).

The decision on what crops to grow will depend on many factors, such as soil, climate, market, and high-hazard species present. These factors usually vary from airport to airport and hence it is impossible to issue strict regulations as to crops. The Canadian Ministry of Transport, which administers all major civil airports, has expressed reluctance to allow agriculture on airports but does not forbid it (95):

> "Generally, where leasing of airport lands for growing of cereal grains, market garden vegetables and other crops attractive to birds is unavoidable, these crops should be kept as far as possible from runways. Where plowing attracts flocks of birds to runway areas, it can be done during the hours of darkness, or during periods when the problem spe-

> cies are away from the airport area, e.g., during nesting season for gulls — early spring and late autumn for migratory species, etc.
>
> Similarly, cultivating and harvesting practices that attract flocks of birds can often be retimed or changed to avoid creating problems."

The MOT guidelines state that areas 1,200 feet and more away from runways may be leased for agricultural purposes, using the following order of preference: (a) hay, (b) alfalfa, (c) flax, (d) soy beans, (e) pasture (golf courses, driving ranges, etc. within the securely fenced area), (f) fall rye, (g) fall wheat, (h) spring wheat, (i) barley, (j) other cereal grains except corn and oats.

The GAF manual recommends that agricultural utilization of airfields should be avoided because the growing of crops attracts birds, especially during plowing, sowing, and harvesting (201).

When deciding what crops, if any, to allow at an airport, an important consideration would be the amount of cooperation that can be expected from the farmer to help keep down bird numbers (e.g., by carrying a shotgun on his tractor during harvest).

WASTELAND AND WOODLOTS. Wasteland is all the area that is not built up, paved, pastured, cultivated, or covered by trees and water. At most airports there is little wasteland, but wherever the ground is not utilized by man, birds make good use of it. Wasteland usually has many tall weeds, grasses, and shrubs that attract birds. In addition, there are often peculiar features, such as an unused shed, a deserted farmhouse, an old windbreak, the remains of a piggery or a market garden, a few tall trees, rotten fence posts, etc. Such attractions have to be removed.

The bushy vegetation itself should be kept short, or be replaced by grass, crops or other plants. Depending on local conditions, the ground cover can be cut, burned, killed with herbicides, or bulldozed.

Woodlots are parcels of tree-covered land on airport property. Generally speaking, their presence provides many nesting, resting, roosting, and feeding opportunities for birds. In addition, they provide "edge effect," the phenomenon that the borderline of two types of biotope (e.g., grass and trees) is usually rich in bird life. Where it is not permitted to cut the trees, all undergrowth should be removed, and the trees should be frequently inspected for the establishment of colonies of nesting birds (crows, rooks) or of roosts (starlings). It is not likely that a rook colony on the airport will remain unnoticed, but starlings may come and go to their roosts with few people being aware of it, because they arrive at dusk and leave at dawn. This was the situation of Airbase Lahr, West Germany, where starlings roosted by the thousands in a large stand of dense, tall shrubbery that was separated from the runway only by a narrow strip of grass (51). Clearing of this shrubbery solved this problem.

OPEN WATER. Shorelines, marshes, lakes, ponds, pits, creeks, canals, ditches, gullies, wet meadows, pools, and puddles are attractive to a good many species. Birds need water to drink and to bathe. In addition, water-bodies, even temporary ones, often support a large insect fauna.

The most effective way to overcome water-associated bird problems is draining, filling, levelling, and, if need be, pumping. To a large extent all this has been done at Vancouver Airport. Operations were spread over a period of several years, because of the high costs involved. When ditches could not be replaced by culverts, wires were strung across them to discourage ducks and herons. Because of the high ground-water levels, pumping was often necessary during high tide. The results have been highly satisfactory, but the bird problem at Vancouver Airport is not completely solved because it has many other attractive features (192). At several other Canadian airports, notably Montreal, Toronto (175), and Winnipeg (191), similar operations, though on a smaller scale, substantially reduced the availability of open water and resulted in smaller bird numbers on the field.

There may be some controversy regarding the most desirable grass length, but there is general agreement that standing water is an amenity for birds, and all manuals and instruction pamphlets recommend the airfield be drained and open water eliminated as much as possible.

BUILT-UP AREAS. Sparrows, starlings, pigeons, gulls, and crows have adapted themselves to the ways of man and make use of many opportunities provided by him. The built-up area of an airport offers "city birds" a variety of nesting and resting sites, and — with luck — a few bites to eat as well.

Birds in and around buildings are not only a nuisance because they may become involved in an aircraft collision when moving about, but also because they dirty the hangars. Droppings of birds nesting or roosting in hangars can make a slippery mess and can corrode the skin of aircraft. When an aircraft is opened for overhaul, bird droppings, feathers, and nesting materials may foul the electronic equipment. It is also possible that birds that have just fledged may get stuck somewhere inside the aircraft, and their dead bodies might cause more than a bad smell.

If birds are nesting or roosting on top of hangars and other airport buildings they are much less of a pest, but they may deface roofs and facades with their droppings. This type of fouling may become a sanitary nuisance if it remains unchecked (63).

An obvious way to keep birds out of hangars is to prevent them from coming in. Hangars have many holes and openings that birds use as an entrance, and access can be prevented by screening. But in active hangars the doors are opened frequently, and even when they are closed usually some space above or below them allows birds to get in.

When it is impossible to keep bothersome birds out of a hangar, it may be possible to remove or destroy the attractions that draw them inside. Favourite spots for nesting can be screened or covered with wire. As discussed earlier, frequently used perches can be made unattractive with wire or with a soft sticky material that can be applied with a caulking gun. The birds dislike the feeling of this gel-like material under their feet and will leave the treated perches. This method would work best in a hangar with relatively few places for the birds to sit. Use of strobing lights and poisoned perches in hangars have also been discussed earlier in this chapter.

Against birds nesting and roosting on the outside of buildings, basically the same methods can be used.

LANDSCAPED AREAS. Ornamental shrubs and trees often bear colourful flowers, which attract insects and hence birds, or seeds and berries that are eaten by birds. When the landscaping of an airport is planned, factors such as appearance, soil, water level, climate, costs, and maintenance are of prime concern, but birds should also be considered. Expert advice may be obtained from biologists with the Department of Agriculture or the nearest university.

The GAF has published a list of woody plants that should not be used at airports because of their attractiveness to birds. They also issued a list of shrubs and trees that are acceptable in relation to bird hazards. Their instruction manual also recommends that plantings should never be done in the form of hedges or thickets (which are ideal for foraging and nesting), but of repeatedly staggered lines of different species (201).

The Canadian Ministry of Transport issued a list of trees and shrubs selected for airport planting. Those attracting birds are so marked, and "their use should be carefully considered on site by site basis" (96).

MISCELLANEOUS BIRD ATTRACTIONS. Garbage dumps, perching sites, and unusual nesting sites are bird attractions that are not directly associated with types of terrain.

By and large, garbage dumps are a major source of food for high-hazard birds. Waste disposal areas are often an eyesore, and sometimes also a health hazard. Waste disposal on the airport should not be allowed, and removal of a dump should be the first goal of any environmental approach to airport bird problems. Where off-airport garbage disposal is impossible, edible refuse should be contained in heavy plastic bags and all garbage should be buried with earth immediately after being dumped; disposal operations would also attract fewer birds if done at night. Several airports (e.g., Vancouver Airport, 192) have eliminated their dump, and good results were obtained in each case.

Perching sites, preferably a variety of them, are prime requirements for birds. Perches mostly consist of some vantage point rising above the surroundings from a matter of inches to 50 feet or more. These perches are used by birds for singing and calling and display, as observation points and hunting lookouts, as spots for resting, digestion, preening, roosting, and for social gatherings. Lone trees, hedgerows, fences, gates, posts, shrubs, stumps, junk, weed-patches, and boulders should be removed. Power lines are a well-used perching facility and, ideally, they should be put underground. Runway approach guideposts and runway lights should be equipped with sharp spikes to make them useless as perches (175).

Birds have an uncanny eye for nesting opportunities. At airports they may nest under bridges, in culverts, in old shacks and work huts, and on radar towers. At Leningrad Airport, USSR, for instance, a Great Spotted Woodpecker pecked a hole through the foam plastic covering of a radar antenna and brought in nesting material (232). Unused sheds, etc., should be razed or boarded up. Nesting elsewhere can be prevented by screening with netting or wire.

Aircraft parked on the apron or on the field have also been used for nesting. Birds usually enter the aircraft through some small access holes just large enough for the species concerned. This makes it difficult to remove the nests, because much work is involved in removing access panels or whole sub-assemblies.

At least 29 nests in aircraft involving 8 bird species were described in one report. Nests were found in the process of being built, with eggs, with young, and long after the birds had left them (69). The longer an aircraft is out of active service, the more likely it is that the parents can raise their brood. Birds can build their nests in an aircraft parked for a weekend, or even a single day. From the USSR it is reported that starlings began bringing nesting material into an opening in a AN-10 aircraft as soon as it had come to a stop after landing (232). At Montreal Airport, starlings placed within an hour so much nesting material in the inlet of a Vanguard that engine shutdown and component replacement were required (374).

Two North Star (DC-4) aircraft that had been parked for a long time on the infield at Toronto Airport were "liberally sprinkled" with starling nests, the birds flying back and forth across runways and taxiways. It was recommended that these aircraft be moved farther away from the operational runways, and that parked aircraft, out of service for a short period, be checked at least twice a week from April through June (175).

ALTERNATIVE GROUND COVER. The techniques discussed so far are not novel. The suggestion of using an alternative ground cover goes a step further in that it proposes the replacement of existing vegetation by a specially selected ground cover, other than grass or crops.

As mentioned earlier, a healthy turf is generally used at airports, especially around all paved areas to hold the soil, to improve drainage, to provide support for aircraft that run off the paved surface, and for crash rescue vehicles and fire trucks. The pleasing appearance, moderate maintenance requirements, and its economic use are further advantages of having grass at airports.

Obviously, any other plant or material with the same "qualifications" as grass, but with no or almost no attraction for birds, would be even more suitable for use at airports.

So far, little work has been done to determine what plants would be best suited for use as "anti-bird" ground cover on airports, and the use of an inorganic material has merely been suggested now and then at flight safety meetings.

Looking first at the possibility of growing a selected plant as anti-bird ground cover, such vegetation would have to meet many criteria (22, 182). It would need to be:

- unattractive to birds, either as food or as cover;
- unattractive, either as food or as cover, to invertebrates and other animals preyed on by birds;
- of a low, spreading form of growth, not exceeding 10 inches in height (to allow passage of emergency vehicles);
- without shade requirements, and adapted to open areas;
- aggressive towards invading plant species, and dominant;
- quickly established and practical to maintain;
- able to prevent soil erosion;
- capable of supporting emergency vehicles, and, as much as possible, aircraft;
- not a fire hazard.

With so many conditions to be met, it will come as no surprise that research and development work has been slow. Candidate plant species should be given a fair chance to prove themselves, with careful monitoring of their health and performance for several years.

In Europe, the Dutch tested use of lucerne instead of grass (421), but they were probably unsuccessful because no further reports were published.

In Canada, a study was carried out at Vancouver Airport and more comprehensive tests were made at various places in eastern Canada. At Vancouver, nine plants species, including two grasses, were tested on small plots on the airfield. It was concluded that none of these could successfully compete with the well-established native grasses and weeds. However, these studies were of a preliminary nature only (182).

The eastern Canada tests began with six native plant species: three-toothed cinquefoil, black medick, low hop-clover, broom-crowberry, silverweed cinquefoil, and mouse-ear hawkweed. (No exotic species were considered because the introduction of aggressive, non-native species may develop into a costly agricultural weed.) These six species were chosen because they appeared as aggressive components in the vegetative cover on or about various eastern Canadian airports, and in fact, already served as suitable ground cover in some airport areas.

The first four species appeared unsuitable as alternative ground cover for reasons of slow growth, lack of aggressiveness, lack of winter hardiness, and flammability. A fifth species, silverweed cinquefoil, might be suitable, but only under certain conditions of soil moisture. The remaining plant, mouse-ear hawkweed, showed good promise as ground cover on airports with infertile soil and further experiments are now being carried out with this species on a four-acre plot at CFB Summerside, Prince Edward Island.

So far, these long-range studies have illustrated the great difficulty in obtaining enough seed of a plant with no commercial use, and of finding the proper seeding procedures to grow one particular plant species on a small field. The maintenance of such a pure stand (monoculture) would probably be considerably more difficult than that of the grass species plus a few invading weed plants. Monocultures are much more liable to disturbances (insects, severe weather, invading plants, etc.) than multi-species plant communities. This means that the maintenance of the new ground cover would likely require a great deal of attention and energy input in terms of fertilizers, herbicides, and pesticides (22).

The experiments have already made it clear that the introduction of any new vegetative ground cover at an airport is not by itself going to solve the local bird problem. Alternative ground cover is only one approach (and a difficult one at that), and will have to be used in combination with other techniques (22).

Apart from plants, inorganic materials might be used to make a bird-free ground cover. Coarse gravel, coke cinder, waste from ore smelters, and similar materials spread over an airfield would certainly reduce local bird numbers, would probably improve the drainage, and might support traffic if tightly packed, perhaps even without much damage to tires. The black cinder field could be spray-painted in an attractive green shade, and liberal applications of herbicides would take care of any natural green that managed to push through. In most western countries, where grass is "socially accepted," this approach would be considered with the greatest reluctance. In addition, a practical problem would be to keep the runways and taxiways clear of gravel and cinders. Such objects are sometimes drawn into engine air intakes, causing "foreign object damage" (FOD).

Artificial turf as used in football stadiums might be more suitable, but costs would be high. Its use was rejected by the Canadian Committee. The difficulty of fastening the turf in place and the effect on it of high-temperature engine blast were considerations in its rejection (374).

A less radical method would be application of chemicals that would bind the soil particles, which would thus prevent wind erosion and hamper seriously the establishment of vegetation. This approach was suggested in Australia (423) and is being considered in the US.

PROVISION OF GREATER ATTRACTIONS OUTSIDE THE AIRPORT.

One way to make the airfield less attractive is to make its surroundings more attractive. The idea that birds can be removed from an airport by offering them greater allurements elsewhere may be misleading. If, for instance, a bird sanctuary were established in the near vicinity of an airfield with a bird problem, many airport birds could probably be encouraged through an intensive shellcracker campaign to move to the sanctuary. Because of its ideal conditions, the sanctuary would soon carry a capacity crowd, and surplus birds would start drifting to the second best habitat (that is, the airport). The same old problem would soon exist, plus a new one involving many birds nearby that might also become a hazard.

Any proposal to create a greater amenity for birds outside the airport should be based on an understanding of what species are a nuisance, and why. To contrive an example, assume that a large garbage dump is operated a few miles from an airport. Food is plentiful, many gulls frequent the dump, all gulls loaf on a paved part of the airfield, and no other loafing areas are available. An alternative loafing area close to the dump is provided, and attempts to scare the gulls from the airport are stepped up. There are now two possibilities. The first is that the gull population does not increase and that the birds start to loaf on the newly created resting area. This is likely to happen when the dump or the nearby roosting area are already serving a maximum number of birds. The other possibility is that the dump can easily feed many more gulls, that the nearby lake can be used by many more birds for roosting, and that the shortage of loafing areas has now been eased by the friendly folks from the airport. More gulls will soon move to this attractive habitat and both loafing areas will eventually be used by resting birds. The point of this artificial example is that knowledge is lacking on the factors that limit the size of the local gull population. If the limiting factor happens to be "loafing area," providing more of it is likely to result in more gulls in the area rather than their disappearance from the airport.

Despite these cautionary remarks, birds can be, and have been, made to give up airport facilities by provision of suitable alternatives elsewhere. The best-known example of this approach is the creation of artificial roosts near Auckland Airport, New Zealand. This airport is largely surrounded by tidal mudflats that are ideal feeding grounds for many waders. During high tide, when the mudflats are not available to the birds, they retreat to "high tide roosts" where they preen and rest. Auckland Airport was used as a high tide roost by thousands of birds, and although they had caused less damage than might have been expected from their numbers, they were considered a serious potential hazard due to their habit of flying in large, tight flocks.

A nearby six-acre salt marsh covered by scrub and rushes was reclaimed with 28,000 cubic yards of filling and developed as an artificial roost (basically a large bare area), complete with pools (for drinking, bathing, and wading) and low windbreaks (for shelter in rough weather). The scheme worked well: waders and gulls roosted on it consistently, and herons and ducks sometimes. There were a few minor snags: the waders were sometimes scared off their new roost by Harrier Hawks that were attracted by an adjoining rush area. It was necessary to begin shooting and trapping these raptors, but a more permanent solution would be to do away with all rough growth near the roost. It was also observed that the windbreaks were perhaps a bit too high (the roosting birds were often edgy, showing signs of insecurity) and that the roost required proper maintenance (345). However, the whole enterprise has been described as an "outstanding success" (88).

At Whiteman AFB, Missouri, US, resident flocks of Greater Prairie Chickens (considered an "endangered species" in that state), are a serious hazard because the main runway had been built across their traditional "booming" grounds. A booming ground is a small piece of land where the male birds strut around during the breeding season to show off their prowess. The same booming area is used by generation after generation. As expected, the chickens could not be scared away with conventional techniques, and being an endangered species, neither could they be killed.

Tests with falcons and hawks in 1972 showed that the chickens could be driven off the airfield, but that they soon returned. As chickens prefer long grass, the grass on the airfield was cut. It was also suggested that alternative booming grounds outside the airbase should be provided. Spoiling their habitat on the base, and providing suitable alternatives off-base, in combination with persistent harassment on the airfield, may cause the chickens to give up their ancestral booming area and establish a new one elsewhere (121, 286).

The "lure-away approach" offers a useful solution to a few unusual problems. Costs need not always be prohibitive: at Auckland the roost construction was a welcome project because it provided an easy way to dispose of large amounts of dredged spoil from nearby shallow areas.

AREAS OUTSIDE THE AIRPORT BOUNDARY

This chapter deals with methods to prevent bird strikes at or over airfields, but to get rid of birds at a certain airport it is often necessary to remove an attraction outside the airport boundary. The next chapter describes ways and means of preventing bird strikes that occur away from airports, especially during climb-out and descent. The distinction between these two classes of bird strikes helps to give a clearer picture of the different aspects of the problem. Here, however, it causes some overlap. Gulls loafing on the runway may cause an "at-or-over airport strike," but when flying to a nearby garbage dump they may become involved in an "away-from-airport strike." Thus, closing the garbage dump would probably reduce the incidence of both types of strikes. This section briefly mentions a few garbage-related airport bird problems, and the next chapter deals with bird concentrations and flight routes outside airports.

Some bird attractions near an airport are difficult to remove, but many others can be largely eliminated, or their appeal can be reduced to a minimum. A seashore cannot be relocated, but garbage dumps, sewage outlets, fish plants, fish piers, abattoirs, and pig farms can be relocated or managed so as to attract fewer birds.

As an airport authority has no direct control over matters outside its territory, it is often extremely difficult to implement the recommendations made by experts. Progress is usually slow for a variety of organizational, legal, financial, and political reasons. It is easy for a visiting biologist to recommend that the garbage dump at the end of the main runway should be closed, but it is a different matter for the airport manager to apply that advice.

Although changes in habitat and commercial activities outside the airport require lengthy negotiations, it has been evident that where good proposals have been followed through, significant improvement in the bird situation has resulted.

A case history of a Canadian airport makes this clear. Up to 1960, Lakehead Airport (now Thunder Bay), which is located about three miles from the shore of Lake Superior, had only a minor bird hazard problem, mostly during the autumn migration. In 1960 the city of Fort William closed the municipal garbage incineration plant and opened a sanitary landfill area adjacent to Lakehead Airport property. Immediately, Herring Gulls began to frequent the airfield, and despite intensive efforts during a 2½-year period the gulls could not be driven off the airport. Often as many as 2,000 gulls were seen on airport property. A representation to the Fort William Council by the Bird Hazard Committee resulted in the decision to discontinue the landfill and to return to the incineration of municipal garbage. As soon as the dump was closed, gull numbers at the airport decreased greatly and the bird hazard was again at an acceptable level (116).

Windsor, Ontario, may serve as an example of how such things should work: when the municipal landfill area was full, the city considered several new sites, one of which was next to the airport. The Bird Hazard Committee, called on for advice, recommended that the site farthest from the airport and close to a river be developed. Despite higher costs associated with this more remote area, the city followed the recommendation and a serious bird problem was prevented. In contrast to these favourable outcomes, some airports in Canada have been unable to arrange the removal of a nearby garbage dump, even after several years of effort.

In other countries also, the practice of dumping garbage has created serious problems. A huge dump of the city of Marseilles in southern France was used by large numbers of Herring Gulls and Black-headed Gulls. Their daily flights, monitored by radar, between the dump and their favourite roosting ponds were a serious concern to nearby military and civil airports (257). Near Sydney Airport, Australia, the closing of one garbage dump, and the change at another to dumping and covering at night, reduced local gull numbers from over 10,000 to less than 2,000 (423).

The problem of gulls, garbage, and airports is probably nowhere more complex than in the San Francisco Bay area in the US, with more than 30 solid-waste disposal sites, 7 major and 13 minor airports, and a wintering gull population of about 140,000. Experiments with marked birds showed that individual gulls were not restricted to a particular dump in the area, but often interchanged. This means that any appreciable reduction of the overall hazard will require regional or at least subregional control of the solid-waste sites (111, 112).

PLANNING OF NEW AIRPORTS

All too often, airports have been built at great cost without any anticipation of bird problems. As an afterthought, strike hazard reduction schemes at disproportionally high cost are reluctantly initiated once the bird hazard has become more than a minor nuisance.

Minimizing bird problems should begin concurrently with site selection for the new airport. Airports are most needed in the overcrowded areas where large tracts of open land no longer exist or are too costly. New York City has been considering the creation of a $7 billion "airport island" some five miles out to sea as an alternative to a $5 billion airport inland, which would cause problems of ground transportation, noise, and pollution (15). Needless to say, gulls would welcome the development of a large loafing area at sea and would probably cause a persistent problem to air traffic.

In Denmark, a natural island (Saltholm) about five miles offshore from Copenhagen Airport, has been considered as the site for a new Copenhagen II Airport (127). This flat, virtually treeless island has a few farm buildings, some free-roaming cattle, a large Herring Gull breeding colony as mentioned earlier and large numbers of wintering waterfowl in the shallow water surrounding the island. From the bird hazard point of view this is the last place to build a new airport. In fact, local conservationists campaigned to develop this island as a bird reserve and strongly objected to the planned jetport.

A similar situation exists in England, where proposals to build London's third airport (Maplin) on reclaimed mudflats off Foulness Island caused alarm among conservation-minded people. Foulness Island is located east of London in the Thames estuary. The island and surrounding mudflats harbour an abundance of gulls, waders, ducks, and passerines, most of which are transients. Up to 10,000 dark-breasted Brant Geese (two-fifths of the world population) arrive at Foulness when returning to their winter quarters, and at least half of these remain to feed on their favourite eel-grass, which grows as profusely in only a few other areas in Europe. As some of the alternative feeding grounds will likely be lost due to reclamation, the survival of this goose will become even more dependent on Foulness Island as a wintering area (66).

In the public debate on development of Maplin Airport, experts testified on the likely environmental impact, on the estimated bird strike hazard, and on the costs of maintaining an effective bird-scaring team at Maplin (73). In addition, ecological studies were commissioned to obtain basic data on bird life at Foulness (so that the environmental impact, as reflected by changes in bird numbers, may be measured if the airport is built), and to find methods for minimizing the environmental impact (e.g., by providing suitable habitat elsewhere, especially for the Brant Geese). The latter studies are mainly centred on the problem of determining the ecological requirements of the bird species that may be affected. However, in view of revised estimates of the predicted growth and pattern of air traffic, the proposal to build the airport may be dropped.

In Canada, plans for a new Toronto II Airport were opposed by citizen groups (representing the interests of people near Pickering, Ontario, where the federal government proposed to build the airport) as well as by environmentalists. Because the objectors claimed that the location of the new airport would make it particularly unsafe because of problems with resident and migrant birds, the government awarded a one-quarter million dollar research contract to investigate these assertions. The main conclusion from a year-round study by a team of scientists was that bird hazards to aircraft would be no higher at Pickering than elsewhere in the Toronto area. The potentially dangerous migration of waterfowl, including flocks of Canada Geese, was mainly of a "broad front" nature and not particularly heavy over Pickering (267).

The new Montreal II Airport (Mirabel) opened in 1975. A long-term ecological study of the proposed airport site was begun following the federal government's announcement of the new airport. The investigations dealt with distribution and numbers of birds at and around the site. Of 167 bird species living in the airport area, 38 were considered a flight safety hazard. Bird hazards at Mirabel will be minimized by the "anti-bird awareness" that existed during design and development of the airport, as detailed in the report of the biologist (287).

Based on the above and other studies the Canadian Air Transportation Administration issued planning guidelines for the use of land outside airport property boundary. These guidelines (given in Appendix 5-1) describe the types of activities, commercial and otherwise, that are acceptable in this area, and those that are not.

Although individual countries have begun to draft regulations regarding land use at and around airports, the ICAO has been slow in pushing for international legislation in this matter, despite suggestions to do so (88). International law stipulates the technical requirements of airfields, but at present no laws govern their biological environment.

CONCLUSIONS

The aim of all ecological methods applied to the bird problem is to make an airport less attractive to local nuisance species, without introducing new bird problems. Most of the techniques described reduce bird numbers without creating attractions for other species. With other control methods this might not be true. If, for instance, all perches for hawks (big trees, high posts, etc.) are removed from an airfield, it may become more attractive to the birds on which they prey, but to what extent this may be true is difficult to determine.

The number and variety of birds is reduced when an airport is made uniformly unattractive to them, but they are not eliminated completely, if only because the runway, which cannot be altered, is used by gulls and waders for loafing and resting. Major airports located in a bird-rich area will continue to have a bird problem even when all possible habitat changes have been implemented. However, this residual problem will be relatively small, and the remaining birds can be driven off more easily because fewer attractions are available and because effective dispersal techniques can be, and have been, developed. In short, a bird dispersal team, or at least some bird-scaring capability, will still be needed for maximal flight safety.

Some habitat changes, such as the removal of waste disposal sites and improvement of drainage, are effective and can be implemented without advice from bird experts; for other modifications, such as crop selection, biologists and agricultural specialists should be consulted.

The most common obstacle to any ecological approach is lack of funds, or to be more accurate, a lack of proof that costly changes in the airport habitat will pay in the long run. The Canadian Government has spent millions of dollars over a period of several years to make its major airports less attractive to birds, and the best way to judge the cost effectiveness of these operations is to examine the bird strike statistics for Canadian airports. As explained in Chapter 2, bird strike statistics are often unreliable and difficult to compare, but Air Canada has paid a great deal of attention to this problem and its personnel have been reporting bird strikes consistently for many years. As most strikes on Air Canada aircraft occur during takeoff and landing, most damage will occur at or over airfields. Thus a reduction in bird strike damage over the years may be taken as evidence that airports have become safer. Air Canada's damage for flights inside Canada do show such a trend: average yearly damage decreased from $173,000 during 1959-63, to $116,000 for 1964-68, and to $86,000 for 1969-73 (Appendix 2-3, and 47). Although these statistics are not strictly comparable, because of changes in fleet size, type of aircraft used, routes, inflation, etc., they have been used to support the claim that bird hazards at Canadian airports have been noticeably reduced.

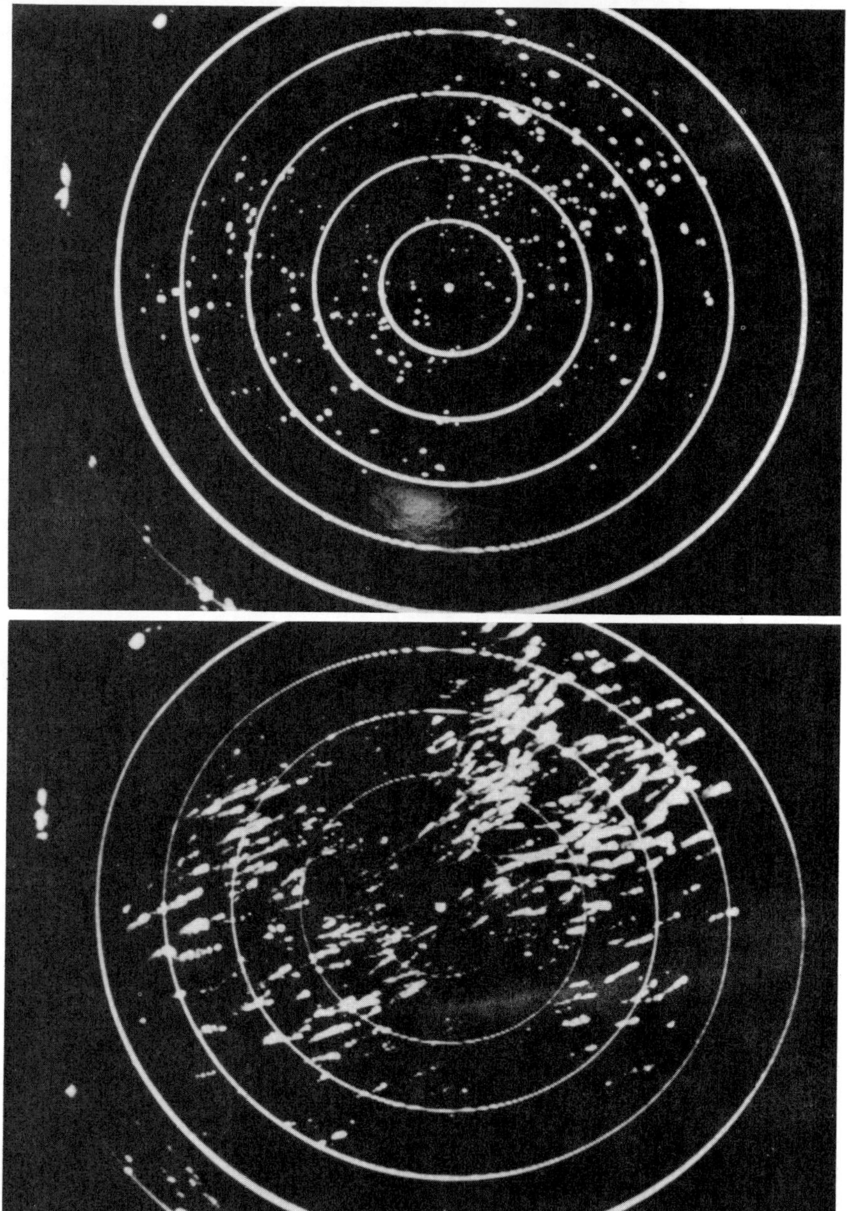

Radar "echoes" of flocks of migrating Snow Geese on the screen of an AASR-1 surveillance radar at Winnipeg International Airport, Canada. The range rings are 10 nautical miles apart, and the geese are flying in a north-easterly direction. (top) Normal presentation (where echoes fade away after a few seconds) showing the bird echoes as bright dots. (bottom) Bright scan display (where echoes persist for several minutes) showing the echoes as bars.

Chapter six: Give birds the right of way

PREVENTION OF BIRD STRIKES AWAY FROM AIRPORTS

Pilots in modern aircraft moving at high speed and birds flying nearby have little time to avoid each other. Pilots tell how a tiny distant speck in the sky almost instantly becomes a big bird whirling past. Furthermore, pilots do a lot of "office work" (reading maps, checking instruments, keeping records) during flight and cannot spend all their time watching for birds (or other aircraft, for that matter). Many smaller birds pass unseen, and many strikes go unnoticed until after the aircraft has landed.

If pilots cannot see the birds, then ground personnel should tell them where the birds are flying. This is the rationale behind various programs to draw up bird maps, to predict bird migration, and to develop bird radars.

This chapter reviews the methods followed to provide pilots and air traffic controllers with useful information on bird hazards during flight. Before doing so, it briefly discusses the options available in dealing with situations when dangerously large numbers of birds are on the wing.

PROCEDURES TO MINIMIZE STRIKE RISKS DURING PERIODS OF HIGH BIRD DENSITIES

Reduction of bird strike risks is based on avoidance of the birds, and "softening the blow" when they cannot be avoided. Pilots are often unable to avoid birds by manoeuvring around or away from them. It is only pilots of light aircraft, flying at low speeds in daylight hours, who have a good chance to see and avoid their feathered counterparts.

AVOIDANCE OF BIRDS

The main alternative courses are summarized here:

- Plan a flight to avoid areas of high bird density, as much as possible. This applies especially to military fighter aircraft that fly low at high speed, but that often have the option of using alternative routes and areas for training flights.

- Travel as much as possible above the "bird layer." Few birds fly over 10,000 ft. above ground level, and most are below 5,000 ft. This applies both to commercial aircraft and to military transports. In practice it means that climb-out and descent should be as steep as is safe and operationally feasible.

- Change runways, or delay takeoffs and landings. These procedures are useful when the air traffic control radar shows bird hazards are localized or of brief duration.

- Close the airfield and divert incoming aircraft to other airports. This drastic step can be taken when the density of airborne birds is so heavy that the bird strike risk becomes unacceptably high.

The object of the above procedures is for the aircraft to avoid the birds. In addition, there are two actions that may help the birds to avoid the aircraft.

- Have landing lights on at all times when flying below 10,000 ft. This gives the bird a better chance to see and more time to avoid airplanes, as discussed in Chapter 4.

- Descend and climb-out in a straight line. This makes it easier for the birds to anticipate the flight path of the aircraft and thus to get out of its way, as mentioned in Chapter 1.

"SOFTENING THE BLOW"

For this option, there are three possibilities:

- Reduce speed. The force of impact when a bird collides with an aircraft is roughly proportional to the square of the aircraft speed, as detailed in Chapter 3.

- Fly with helmet visor down. This applies to military aircraft pilots only.

- Duck your head below the level of the windshield when a collision appears inevitable (97).

BIRD DISTRIBUTION MAPS

Birds are not evenly distributed over the land. Some areas have tremendous numbers of birds and others have few. The bird density may vary with the time of day and with the season. As discussed in Chapter 1, countries in the northern hemisphere with a moderate climate have four categories of birds: permanent resident (present in the area year-round); summer visitor (breeding in the area but wintering farther south); winter visitor (breeding farther north but wintering in the area); and transients (breeding north and wintering south of the area). Although the density in an area often fluctuates considerably in the course of a year, the differences from year to year are usually relatively small. This makes it possible to plot bird maps that show high-hazard areas for certain periods of the year.

The information on bird distribution can be plotted on two kinds of maps: small-scale maps that cover large areas (e.g., part or all of a country or continent) and large-scale maps that cover small areas (e.g., an airport's Terminal Control Area). Obviously, large-scale maps are much more detailed than small-scale maps.

SMALL-SCALE BIRD DISTRIBUTION MAPS

In Europe, some countries have prepared small-scale maps. As gulls, waders, and waterfowl are of special concern, most waterbodies and coastal areas are considered high-hazard areas. In Holland, for instance, the Wadden Sea, Yssel Lake, and Delta Area are occupied by hundreds of thousands of birds both in summer and winter (13). Maps showing the main migration routes and greatest gull concentrations in France were issued by the French Ministry of Transport (258, 259). The Directorate of Flight Safety of the RAF distributed maps showing bird concentration areas in the UK and warned "Flying below 1500 feet is not recommended in these areas" (341). Many waterfowl winter in Denmark and their distribution and numbers are monitored by wildlife biologists to plot high-hazard areas (235, 236).

In addition to these maps prepared by and for individual countries, there have been efforts to co-ordinate studies so as to arrive at maps covering most of Western Europe. This was the aim of the Bird Movement Working Group of the Bird Strike Committee Europe (Appendix 7-2). The results were published in the so-called "Flip Book" of the USAF (416), which contains four maps to cover Europe. Presentation of a great deal of information has resulted in maps that might not be readily understood. One of these maps is reproduced as Fig. 6-1, in which the bird concentration areas are largely correct, but the arrows that represent the migration routes are probably an oversimplification. In fact, the idea of narrow, sharply defined migration routes is somewhat obsolete. As detailed in Chapter 1, most species migrate on a broad front, although some daytime migrants may follow coastlines, etc. (so-called "guiding lines").

In North America, concern about the migration routes of big birds, especially waterfowl and cranes, has resulted in the publication of bird maps by Canadian and American federal departments of transport (98, 417). The maps for Canada were prepared with assistance from the Canadian Wildlife Service. The map for the fall migration of Canada Geese, Sandhill Cranes, and ducks through Canada is reproduced as Fig. 6-2. In the US, bird maps were prepared by biologists on behalf of the Federal Aviation Administration (30-32) and distributed by the FAA (417) and the USAF (25).

Panama, in Central America, has a rich resident bird fauna. In addition, many North American migrants moving to and from South America pass across the isthmus. A map with migration routes through Panama was prepared by the USAF in an effort to reduce the bird strike hazard at Howard AFB and the Balboa Test Range in the Canal Zone (26).

Small-scale bird maps can be used in four applications. First, long-term planning of operational and training flights can allow for the hazardous conditions in certain areas during predictable

Fig. 6-1: Bird distribution map for northwest Europe (after 416). ▶

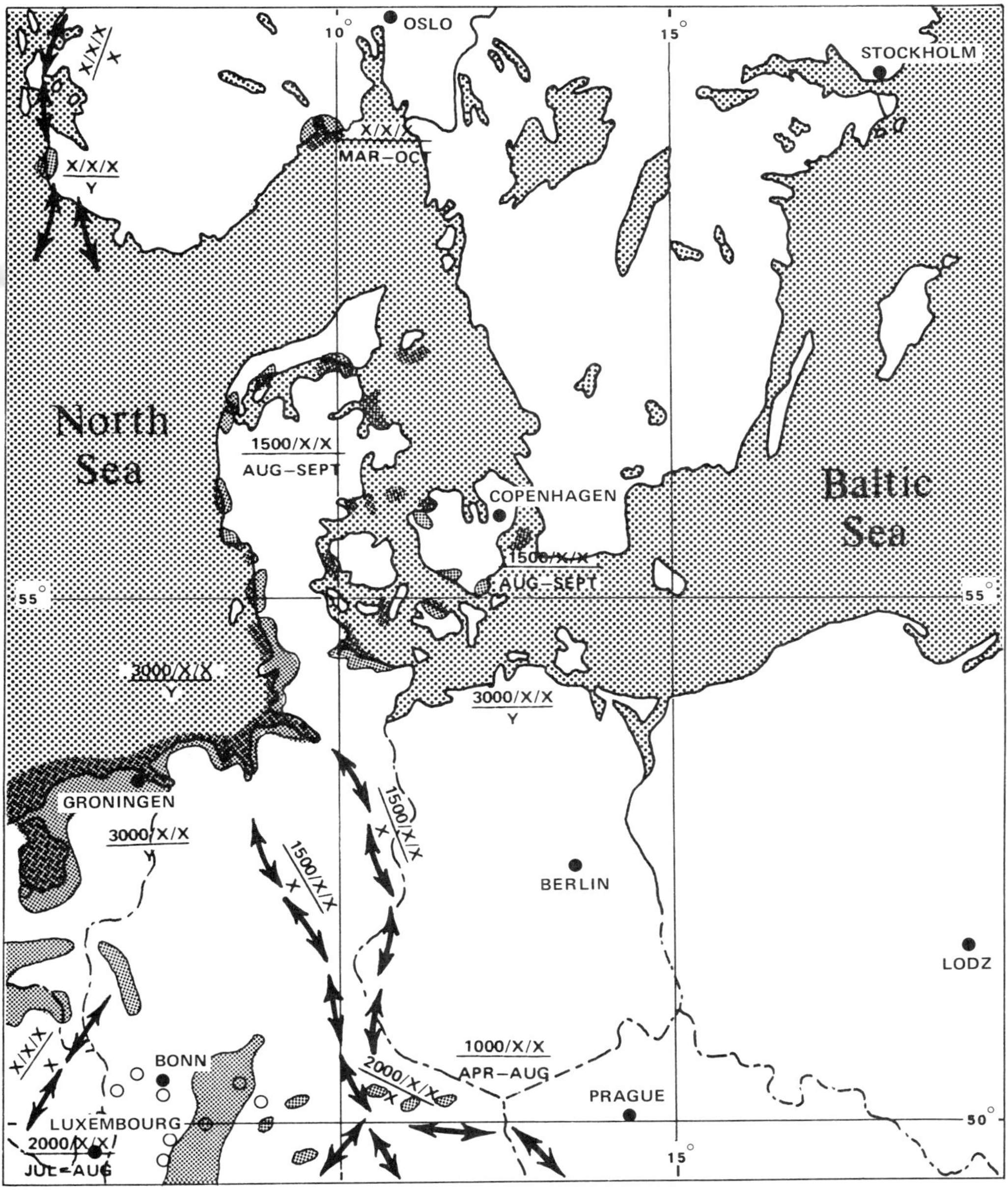

Bird Concentration Areas

Migration Routes

Number of Birds
10 = 10,000 or less
50 = 10,000 to 50,000
100 = 50,000 to 100,000
100+ = Over 100,000

Type— S = Small M = Medium L = Large

Maximum flight level in feet—2,000

Months with highest concentration — JAN-APR.
Y = Permanent

Concentration Data and Migratory Routes = Maximum Alt. / Number / Size over Months of Activity

Example — 1500 / 10 / M over JUN-SEP

Note: Unknown information will be shown with "X"

Example — 1500 / X / M over JUN-SEP

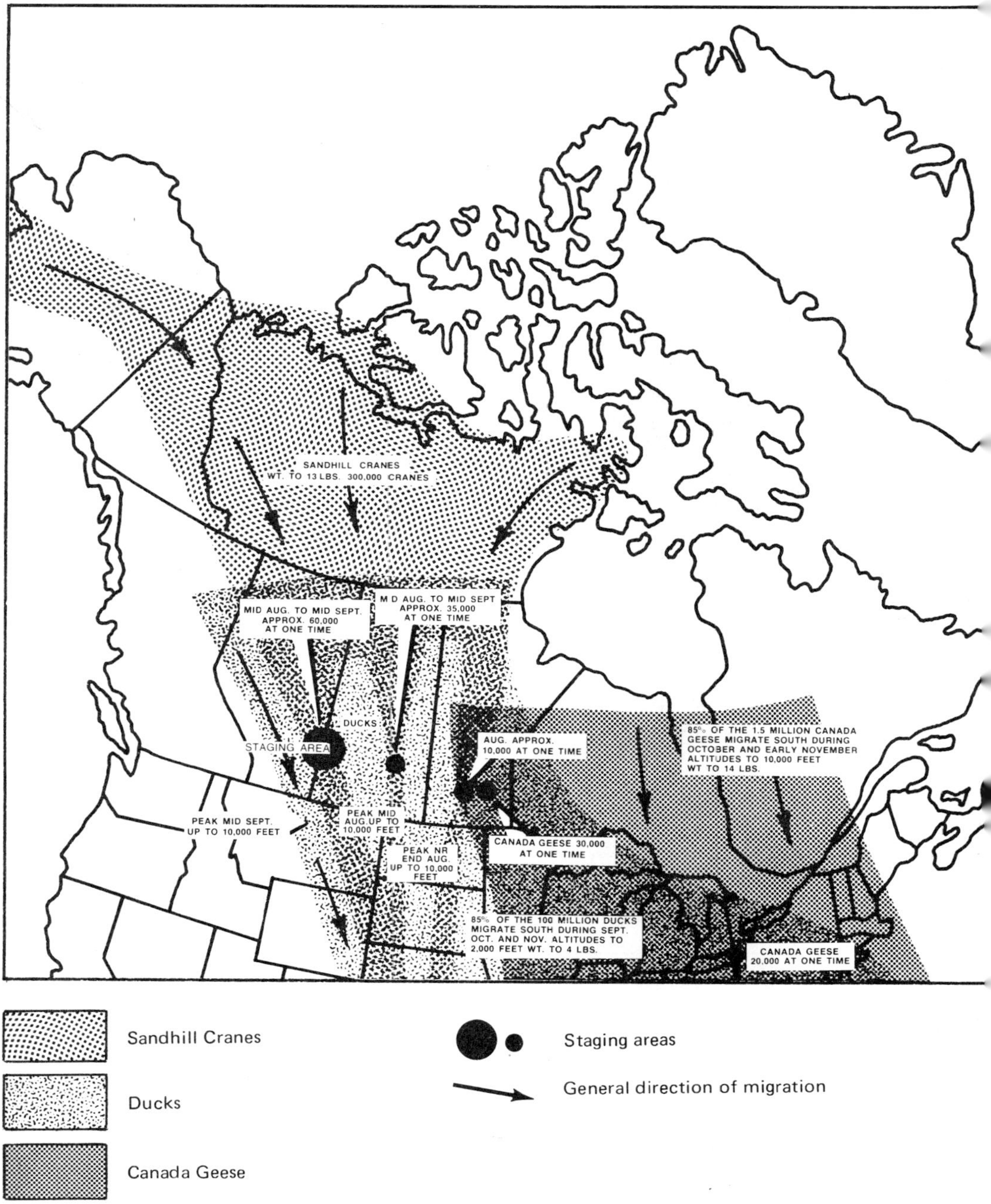

Fig. 6-2: Bird distribution map for Canada, showing fall migration of Sandhill Cranes, Canada Geese, and ducks (after 98).

periods of migration. For example, low-level night-flying exercises in the Cold Lake area of Alberta should not be planned during the last week of August or during September, the period with heaviest migrations (330). Second, for day-to-day flight planning, bird maps can indicate what areas to avoid at stated times of the year. Third, when flying through highly hazardous airspace, the pilot can be forewarned of the bird strike risks, and take the precautions mentioned above. Fourth, bird maps can show aviation experts who are planning new airfields what areas are most likely to have bird problems.

Many small-scale bird maps are rather crude, not only because they cover a large area but also because detailed information is lacking on the distribution of several high-hazard species during their migration. Although most bird migration studies are done to provide biological information, a number of them have been primarily aimed at learning the distribution, numbers, heights, and timing of migrating birds that have caused (or are thought likely to cause) serious collisions with aircraft.

The continuing studies of the Whistling Swan in North America are a good example. After a Viscount crashed near Baltimore as a result of striking a flock of migrating swans, a joint US-Canadian study of their migration routes was started. Swans were trapped, banded, and dyed. A few birds were fitted with a lightweight radio transmitter and were tracked by aircraft and truck. A good deal of information on swan migration was gathered (360, 361).

Each fall, over 150,000 Greater Snow Geese spend a few weeks on their staging grounds near Cap Tourmente, on the St. Lawrence River, some 30 miles downstream from Quebec City, Quebec. As their migration routes between Cap Tourmente and wintering areas on the Atlantic Coast (Virginia, North Carolina) were not well known, radar films were made at Quebec City Airport to determine their route. The geese were not seen to migrate up the St. Lawrence River, and are thus not likely to be a hazard for the busy Montreal Dorval Airport (60).

Another Canadian radar study on goose migration was carried out at Thunder Bay, on the north shore of Lake Superior, where large numbers of Canada Geese and Lesser Snow and Blue Geese pass over during the fall (376). In spring, however, these geese migrate farther to the west and pose a serious flight safety problem for Winnipeg International Airport. Each spring hundreds of thousands of birds cross the Winnipeg area, and a near-disaster collision between a B-737 and migrating geese provided the impetus for a study to monitor and predict the waves of Snow Geese (55, 59).

A radar study of the spring migration of between 30,000 and 75,000 Canada Geese through the Okanagan Valley, in south-central British Columbia, indicated that the geese flew over the Rocky Mountains at 12,000 to 15,000 feet altitude (305).

In Europe the migration of 20,000 to 30,000 Cranes across West Germany was studied extensively by biologists of the German Bird Strike Committee. Migrating Cranes were tracked on radar and observed with binoculars by a network of field observers (199, 242, 243). Crane migration over the Baltic Sea was observed in a Swedish radar study (5). Where needed, warnings were given to air traffic.

Radar experts of the Norwegian Air Force, ornithologists, and flight safety officers have cooperated to map the migration route and resting places of an estimated 10,000 Pink-footed Geese that migrate via the Norwegian coast to their breeding grounds on Spitzbergen (268).

Special studies of raptors involved the Broad-winged Hawk (buzzard) migration around the Great Lakes in Canada (176) and that of the Honey Buzzard near Gibraltar at the southern tip of Spain (215, 217).

The information that becomes available from studies such as these can be used to improve and refine the small-scale bird distribution maps, and also to focus attention on certain annual, and largely predictable, bird movements that present unusually high risks to flight safety.

Some of these studies were the immediate results of costly collisions. There is, however, no need to wait till disasters have occurred and people have been killed. Expenditures to study the distribution, chronology, height, and weather dependence of high-hazard species seem well justified. Likely candidate species would be the Sandhill Cranes, Lesser and Greater Snow Geese, and Red-throated Loons in North America, and the Bewick's and Whooper Swans in Europe. (As mentioned in Chapter 2, a fatal crash near Ireland may have been caused by migrating swans.)

LARGE-SCALE BIRD DISTRIBUTION MAPS

In addition to small-scale bird maps, it is also important to have large-scale maps that show the Terminal Control Area of individual airports. Such bird maps would be based on visual and radar information. Many cities have local gull and starling populations (at least during part of the year) that make predictable feeding and roosting flights. The crash at Boston, where an Electra flew into a flock of starlings, and at Atlanta, where a Learjet struck a flock of cowbirds, are evidence that local bird flights in the vicinity of an airport must be considered seriously.

Many airports have been built at the periphery of larger cities. Refuse dumps are also frequently located at the outskirts of cities to reduce the costs of hauling. Furthermore, many of the bigger cities are built near or on seashores, rivers or other water bodies. Consequently, they offer ample habitat for wintering gulls that commute between feeding, loafing, and roosting areas.

The surroundings of Heathrow Airport (London's principal airport) include several large open-water reservoirs used as roosting places by an estimated 350,000 gulls. Fortunately, "the main trends of dispersal can be seen with radar to lead away from the airport" (65).

Local movements of Herring Gulls and Black-headed Gulls were studied in the Marseilles area, in southern France, using both radar and field observations (432). In the winter many gulls spend the night on a bay, fly to the Marseilles garbage dump in the morning to feed and rest, and return to the bay in the evening. These "commuter" flights, at a height of 450 feet or more, cover some 17 miles either way and usually last for less than 40 minutes. Departure and arrival are largely dependent on the times of sunset and sunrise. Up to 60,000 gulls have been seen assembled in this area (260). When spring comes, many gulls make daily flights to the Rion Islands in the Mediterranean (a few miles off the coast), which have gull breeding colonies. The movements of these gulls at Marseilles have led to several strikes.

Herring Gulls are the main nuisance birds at Copenhagen Airport, Denmark. They have a large breeding colony area on nearby Saltholm Island. Garbage dumps around Copenhagen attract the gulls. In transit between dumps and colony, they also visit the airfield to loaf or rest (271).

A study of daytime movements of the Silver Gull at Sydney Airport, Australia, showed that the flight pattern was related to localized sources of food and shelter. Peak movements (some of them across the runway) occurred at dawn and dusk when the gulls flew between roosting and feeding areas, but large flights also occurred before and after low tide when gulls flew to and from the beach to feed on intertidal animals (425).

In North America, local gull movements were studied in and around Calgary, Alberta (433), Toronto, Ontario (267), Charleston, South Carolina (151), and the San Francisco Bay area (111, 112), among others.

Local feeding flights of starlings leaving their roost in the early morning are readily seen on radar because the birds produce characteristic "ring echoes." The evening flights are usually less conspicuous on the radar because the birds return from the feeding areas at various times in small groups, without the precise timing of the large-scale departures in the morning. The local movements of starlings (like those of gulls) have been reported from many cities in Europe and North America. Starling roost departures were observed by radar in and around London, England, (139, 188), and in France near Toulouse (432) and Metz (129). In Canada, radar films made at several airfields to study bird migration showed "ring" or "arch" echoes near Toronto (335), London (335), Montreal (130), Calgary (57) and Edmonton (131).

Other bird species may be involved in local flights. The erratic movements of the Bean Geese that roost during fall and early winter on a small lake about a mile from the main runway at Sturup Airport in southern Sweden are a serious hazard. During the day when the geese feed and loaf in nearby fields, it is not uncommon to see a group of 50 to 100 birds flying over the airfield and its vicinity (240).

A Candian air weapons test range straddles the Alberta-Saskatchewan border, and Canada's largest colony of White Pelicans is located on an island only a few miles from the centre of operations. Once the young are hatched, the parent birds make long feeding flights to nearby lakes. These flights are high enough to be detected by a surveillance radar some 40 miles south (52).

These examples show that personnel of major airports should be aware of the possibility that, at certain times of the year and of the day, local bird movements may pose a serious flight safety problem. Airports should have large-scale bird maps at their disposal for use in planning aircraft operations, and in the control of air traffic. Because of the short duration of most local bird flights, incoming and outgoing airplanes could be delayed at times when they could not be safely guided past the hazard area. Although the bird maps would indicate the areas and times of high bird hazard in a general way, the local control tower radar might be able to pinpoint the exact location of hazardous bird movement at a particular day or hour. Radar observations might lead to use of a safer runway (i.e., safer descent or climb-out area), and would show when to resume delayed operations.

BIRD MIGRATION FORECASTS AND WARNINGS

LONG-RANGE FORECASTS

Bird maps that show the hazard of migrating birds record the period during which flights are most likely, but the exact dates and hours are, of course, not given. Migration in temperate regions usually occurs in waves (as noted in Chapter 1), and it is desirable to be able to predict such waves. As migration density depends to a large extent on the weather, it seemed logical to develop a migration forecast system based on these density-weather relations.

Long-range migration forecasts have been made in a few countries. In West Germany, for instance, routine predictions are issued by the Office of Military Geophysics of the GAF every 14 days. They are based on the expected weather development, the progress of migration to date, and experience (205). In Holland the RNethAF has issued flight safety posters immediately before the migration season, to warn pilots that bird movements are to be expected under certain weather conditions. These long-range forecasts serve to alert pilots, air traffic controllers, and airfield crews. However, to be effective in control tower operations, more detailed information regarding the actual bird situation is required on a day-to-day basis.

SHORT-RANGE FORECASTS

Short-range migration forecasts have a higher degree of precision, are valid for a shorter period, and if updated frequently are more useful than long-range forecasts. They are based on known patterns of migration, extent of current migration, weather-density correlations, and the weather forecast (181).

Migration forecasts are issued routinely at CFB Cold Lake, Alberta, located at the northern edge of the Canadian prairies. The Bird Hazards Officer prepares two forecasts each day: the early morning forecast deals with the daytime migration, and that in the late afternoon covers the nighttime migration, which is usually much heavier. The forecasts are presented at pilot briefings. Direction and speed of the wind (at ground level, 3,000 and 5,000 feet) and precipitation are the weather factors used for the bird forecasts. Radar films are made to follow development of the migration and to check the accuracy of the predictions. When the forecaster predicts a certain migration density, flying operations are sometimes cancelled. Although the strike rate has been reduced, it is impossible to say how many bird strikes have been prevented in this way. Strikes are known to have occurred when flying was not cancelled during well-forecast heavy migration periods (54).

The forecast system has provided useful results at Cold Lake, but the system has shortcomings: (1) the density of bird migration is not expressed in terms of bird numbers (per unit of air space) but in numerical categories. A 9-step density scale is used, running from 0 through 8, but the numerical relation between the steps is not known, so that it is impossible to calculate, for example, how much more probable is a strike during density "6" conditions than during density "4"; (2) the forecasts do not provide information on the height of the birds; (3) the need for radar film; (4) the need for trained personnel; (5) dependency on the accuracy of the weather forecast; and (6) the system requires modifications for use elsewhere.

Another Canadian study dealt with the spring migration of an estimated one million Snow Geese. These birds winter along the Gulf of Mexico and breed in the eastern part of the Canadian Arctic. During their spring migration they have a stopover on the prairies of southern Manitoba. When migrating from the stubble fields on the plains to the muddy coasts of Hudson Bay and James Bay, many of the geese cross the Terminal Control Area of Winnipeg Airport. As mentioned earlier, a serious strike involving migrating Snow Geese occurred in spring 1969, and a project was begun soon after this strike to monitor their spring migration in the following years (55), and to develop a method for predicting heavy flights (59). This prediction method was based on records of migration dates for previous years (the geese never flew earlier than May 2 nor later than May 17), the preference of the geese for certain wind conditions, and their avoidance of precipitation. When tested in 1974 under operational conditions, the prediction method yielded poor results, mainly because weather conditions were abnormal and some terms used in the weather forecasts did not fit the migration prediction model (58).

In Europe, the Bird/Weather/Radar Working Group of the Bird Strike Committee Europe is sponsoring several projects to study the relation between weather and bird migration, with the goal of developing methods for predicting migration (368, 375). The weather in western Europe changes quickly, much more so than in central Canada, and it will be harder to determine the key weather factors that initiate waves of heavy migration. From radar studies in Switzerland it was concluded that the behaviour of European birds does not basically differ from that of North American birds, but that the same patterns are less obvious in

Europe due to the faster passage of weather systems (80). Another difference between Canada and most of western Europe is that during the winter the bird population in most parts of inland Canada is low, whereas in Europe many species winter in countries such as Denmark, Holland, and England. If in the course of the winter a severe cold spell occurs, many birds make "weather movements" in search of milder conditions. This means that there is almost no month of the year without migration; this is particularly true in England.

A combination of radar and field studies in Denmark provided information for a forecast system that was tested in fall 1973 (317, 318), and further work is underway.

In Sweden migration studies were carried out jointly by the Swedish Air Force and the University of Lund. A combination of radar and field observations proved successful (5-7) and migration forecasts are planned for the near future (4).

Reports of work in Belgium have dealt with the problems of quantifying radar data (273, 274) and selecting statistical methods to determine the relation between weather and migration.

In France, migration forecasts were made in 1972 but were not repeated in 1973. An extensive study that began in 1968 dealt with the effect of weather on bird migration in southern France, and is soon to be completed (261).

In 1973, Germany was the only European country issuing bird migration forecasts routinely. The Office of Military Geophysics in Bonn used synoptic weather situations for preparing the migration predictions, which were valid for 24 hours (205). This system has been used for several years but, as far as is known, no scientific report has given details or discussed the usefulness of these predictions.

Because migration predictions are based on weather forecasts, the accuracy of the former can be only as good as that of the latter. Furthermore, the few migration forecasts made before 1974 were not detailed, i.e., they did not give the expected number of birds per unit of airspace nor the height distribution. Bird predictions serve to alert pilots and air traffic controllers, but often the only way to deal with highly hazardous migrations is to delay takeoffs and landings. Consequently, the usefulness of predictions has been limited in civil aviation, where almost any delay in the tight flight schedules could annoy passengers and increase operating costs.

In military operations, delays or cancellations of combat training flights are troublesome, because they upset the training schedules. However, at CFB Cold Lake, Alberta, bird hazards to crew and aircraft are considered to be of great importance, and predictions of heavy migration are adequate justification for cancelling low-level flying (183).

WARNINGS OF ACTUAL BIRD MOVEMENTS

It is helpful to know about hazardous situations well ahead of time through bird migration forecasts. However, such predictions can be wrong, and a system that provides warnings of migration actually taking place would be useful. Compared with a forecast, a warning has the virtue of reporting the actual situation, if it is not too late to be of use.

The advantages of radar as a tool to detect and warn of bird movement were recognized by the Canadian Bird Hazard Committee (181) and the military of several European countries.

The theory and practice of radar is a highly technical and specialized field, and is not fully discussed here. Several textbooks deal with radar systems (e.g., 336, 359), and the book *Radar Ornithology* deals specifically with bird detection by radar (138).

Radars have been widely applied, for instance in marine operations (to avoid collision between ships and icebergs, etc.), in weather systems surveillance (to determine the extent and height of clouds and precipitation), in tracking weather balloons (to obtain upper air movement data), and in monitoring air traffic. Different types of radar are used in these various roles. Powerful surveillance radars can detect large flocks of big birds flying at high altitude, at ranges of 80 miles and more. With short-range radars it is possible to track individual birds, bats, and even insects (79, 164, 214, 347). Because surveillance radars are installed at many airports and can cover large areas, they have been the type most used in detection of birds for the purpose of warning aircraft.

Bird warnings based on radar information are issued in Denmark, France, Germany, Holland, Norway, and Sweden. The international exchange of "BIRDTAMS" (coded bird warnings) was organized by the Transmission Working Group of the Bird Strike Committee Europe. When the radar shows that the density of migration has reached a set level, restrictions on flying operations become effective.

In Holland, bird density was estimated on a 9-step scale running from 0 to 8. No, or almost no, bird echoes are present on the screen of a surveillance radar at intensity 0, and the screen is maximally covered by bird echoes at intensity 8. The RNethAF imposed the following minimum flight heights as flying restrictions relative to bird hazards (320):

Intensity	Holland (except 2 ATAF* linkroutes and ranges)	2 ATAF low level area and link-routes	Weapon ranges
0-4	no restrictions	no restrictions	no restrictions
5	1,500 ft.	500 ft.	no restrictions
6	1,500 ft.	500 ft.	300 ft. (with exceptions)
7-8	2,500 ft.	no operations	no operations

* 2nd Allied Tactical Air Force

In Denmark, use of the same method and similar flying restrictions resulted in a statistically significant decrease in bird strikes (351).

The RNorAF started a program in spring 1973 using two surveillance radars in the Trondheim area on Norway's east coast. When intensity 7 or 8 is reached, no flying is permitted below 1,500 feet during the day or below 2,000 feet at night. The migration of some 10,000 Pink-footed Geese through this area is of special concern.

The GAF bases its warnings on radar data (10 Air Traffic Control radar stations are taking routine Polaroid pictures of the radar screen) and visual observations (a network of 1,000 observers). All observations are reported to the central weather office. When more than intensity 3 is reached, a "BIRDTAM" is published indicating "bird movement" (no intensities are given), hazardous areas, height band of the birds, and time of validity. The "BIRDTAMS" are transmitted through the weather and ATC networks. The bird warnings are given to all pilots and low-level flights are not allowed in the area mentioned in the BIRDTAM (except for special cases) (163, 204, 205).

Radars operated by the RAF at Gibraltar are well sited to observe bird movements over the Iberian Peninsula, the Strait of Gibraltar, and Morocco. Both the narrow-front migration of raptors and other soaring birds and the broad-front movements of passerines and waterfowl across the Mediterranean were investigated, and warnings were issued whenever deemed necessary (217).

The Sturup/Malmoe Airport in southern Sweden is not only located in an area rich in resident bird life, but in spring and fall it is also traversed by large numbers of migrants (particularly Wood Pigeons). In the migration season, a mobile military height-finding radar was used near Sturup to provide information on the height of the migrating flocks (241).

In North America there are as yet no formal programs of procedures to use radar in the provision of routine warnings of high-hazard bird concentrations, except at Cold Lake, Alberta. It is noteworthy that radar information is used as standard procedure in the bat avoidance program at Randolph AFB, Texas (227).

Nevertheless, some radar information on bird movements is brought to the attention of flying personnel. The US Federal Aviation Center at Balboa, Panama, for instance, issues NOTAM's (Notice to Airmen) by teletype when large concentrations of birds are detected, especially in the vicinity of airfields (26).

Recommendations to reduce bird strikes at Beale AFB, California, include the suggestion that personnel of Ground Control Approach (GCA) radar should watch for bird echoes, and, when necessary, advise the pilots to break-off the approach and go around (309).

The main shortcoming of most bird warning systems based on radar information is lack of precision in determining the bird density. Ideally this should be expressed as the number of birds per cubic mile, but in the studies mentioned above, density has merely been ranked on an arbitrary scale.

Bird warnings can be used in much the same way as short-range migration forecasts. When many birds are in the air, precautions can be taken, or an airfield in the affected area can be temporarily closed.

In military fighter operations, the restriction on or cancellation of training flights under hazardous bird conditions will disrupt schedules and programs, but will not usually result in insurmountable problems. On the positive side, warning programs may increase flight safety, and as mentioned, the drop in number of strikes in Denmark was statistically significant.

In the field of commercial aviation, where people are conscious of both safety and operating costs, solid arguments will be needed to cancel or delay takeoffs and landings because of a bird hazard. Both birds and thunderstorms are hazards in aviation. An elaborate warning system for thunderstorms is in use, and development of a warning system for birds has been suggested (181). The following section attempts to establish the information that might be required for effective decision-making at the operational level.

AN "IDEAL" SYSTEM TO WARN OF BIRD MOVEMENTS

An operational migration warning system to reduce bird hazards to aircraft away from airfields must be accurate, reliable, practicable, and economical. Ideally, the system should predict the probability of a bird strike per takeoff or landing. (Most of the airliners cruise well above migrating birds, and the strike risk away from the airport is therefore limited largely to climb-out and descent.) Bird strike probabilities can be calculated if the aircraft's frontal area and its angle of climb or descent are known, as well as the average dimensions of the flocks (or individual birds), their height distribution, and their density per unit of airspace. Such calculations have been made in Canada (180, 221, 376) and the UK (349). For example, a strike probability of about 1 in 6,500 takeoffs was calculated for a DC-8 leaving Thunder Bay, Ontario, during a heavy fall migration of Snow and Canada Geese (376). When calculating the strike probabilities it was assumed that the pilot did not try to avoid the birds, nor the birds the aircraft.

In practice, the first assumption is more or less correct. Pilots of airliners are busy during climb-out and descent, and may not notice birds at all. Even if the birds are seen, there is little time or room for avoidance manoeuvres, due to the speed and size of the aircraft. At night, observations are impossible.

Regarding the second assumption, Chapter 1 described how bird behaviour with respect to approaching aircraft varies with the species (waterfowl often try to avoid airplanes, whereas eagles have been reported to attack them). In other words, the second assumption is probably not always correct, and hence the calculated strike probabilities on geese are likely to be slightly overestimated. Until bird behaviour toward approaching aircraft has been studied under a variety of conditions, it will be impossible to make reliable predictions of bird strike probabilities. Relative probabilities are less subject to uncertainty, however. Due to the high speeds at which modern aircraft operate, it may not make much difference whether birds intend to attack an aircraft or to avoid it, because they fly too slowly to carry out either intention. This argument may be true for small, low-flying combat aircraft flying at high speed, but it is not true for big airliners taking off and landing, when their speed is far below their cruising speed.

Despite the lack of knowledge about bird behaviour, it is nevertheless obvious that information is needed on the distribution of birds in the air, which depends on the number of flocks (or birds) per cubic mile and the height distribution within that cubic mile. Birds migrate singly, in loose groups, or in tight flocks (as described in Chapter 1), and their height distribution will vary during the course of a night as well as from night to night, due to changes in the weather or in the species composition of the migrants. If this information was available, it would be possible to calculate not only the probability of a collision with an aircraft (based on the total frontal area of the airplane), but also the chance of a serious strike (based on the frontal area of the air intakes and windshields only).

It is obvious that it is not yet possible to provide the practical airport manager with all the facts he needs before closing his airport because of birds. In the meantime, research is being carried out by various groups to improve the usefulness of existing radars in providing a more complete picture of the bird hazard aloft, and their work is discussed in the following sections.

It is also clear that two separate problems are involved: (1) the relation between radar data and bird movement data, and (2) the relation between bird movement data and bird strike risk. Once these two problems are solved, it will be possible to relate radar data to bird strike risk. Little has been done to study the second problem, which is difficult to solve. It seems justifiable, however, to assume at present that bird behaviour is not important compared to bird numbers, and thus to concentrate efforts on solving the first problem: what can radar tell about the density of bird movements?

THE PROBLEM OF QUANTIFYING RADAR BIRD DATA

When the number of birds in the air is to be estimated from the screen of an ATC (Air Traffic Control) surveillance radar, there are several factors to consider: (1) the characteristics of the radar, (2) the presentation of radar information, (3) the effect of radar "fixes," (4) the height distribution of the birds, and (5) the need for a permanent record of the radar presentation.

The characteristics of the radar

The maximum range at which a specific target (whether crow, glider, or B-747) can be detected depends on the peak power sent out by the transmitter and the sensitivity of the receiver (this combination is often referred to as radar performance), and the radar antenna. The maximum range of detection of a flock of ducks is larger than that of a single gull because the radar cross-section (or echoing area) of the flock is larger. In other words, the bird echoes on the screen (or bird "blips") will thin out at greater ranges because more and more of the smaller birds are not presented on the screen. The range at which this thinning-out begins depends on the characteristics of the radar used.

The resolution of the radar (i.e., the ability to distinguish individual targets that are close together) greatly affects the number of bird blips on the Plan Position Indicator (PPI) screen. A British authority has explained why the PPI echo is dependent on the radar's characteristics and independent of target size: the smallest object that can be resolved or distinguished by a radar is a volume in space called the radar pulse volume or resolution cell. This volume is proportional to the radar range, the transmitted pulse length, and the vertical and horizontal beamwidths of the antenna. For a conventional ATC radar, the resolution cell dimensions may be extremely large. At a radar range of 20 nautical miles, a typical resolution cell volume may be 72 million cubic meters (600 meters high by 300 meters wide by 400 meters long). Within this volume many birds may be spaced out or bunched in flocks, but unless their distribution exceeds the horizontal or vertical dimension they will appear as a single echo. Nothing will indicate whether the cell is occupied by 5 ducks, 50 gulls, or 500 starlings (216).

The presentation of radar information

The PPI has a limited range of brightness: one blackbird at a certain range may produce a barely noticeable echo on the screen, a group of 5 blackbirds may cause a medium-bright echo, and 10 blackbirds (still at the same range) may cause a maximum intensity echo. Consequently, the echo produced by a flock of 1,000 blackbirds would not be any brighter than the 10-bird echo, provided that the 1,000 birds were contained in the same pulse volume (216).

The effect of radar "fixes"

Radar "fixes" are electronic techniques to improve aircraft detection. Modern radars are equipped with a Moving Target Indicator (to eliminate echoes from stationary targets such as power lines, tall buildings, etc.), Sensitivity Time Control (to remove undesired small echoes at close range by reducing the receiver sensitivity), Circular Polarization (to suppress echoes from precipitation and clouds), and other improvements. These "fixes" are often used in different combinations during normal ATC operations and some of them have an effect on the number of bird echoes that will show on the screen (331).

The height distribution of the birds

A radar beam skimming the surface of the earth at short ranges gradually loses contact at greater ranges because of the curvature of the earth's surface, and extends at ever higher altitudes above the ground. All targets below the radar beam (or below the radar horizon, as it is called) remain undetected. On a night when all birds fly near the ground, the range at which bird echoes will show on the radar screen is shorter than when the same number of birds are flying higher. As already noted, the height of migrants varies during a single night as well as from night to night. Bird heights cannot be determined reliably with surveillance radar.

Obtaining a permanent record of the radar presentation

Counting bird echoes on a live radar screen is usually impossible, because large numbers of bird blips are sometimes involved; bird blips are often difficult to discern, or to distinguish from other echoes on the screen; and the situation is usually changing continually. Thus there is a need for a permanent record of the radar screen that allows later analysis.

One way to obtain a record of the display is to photograph it, either with a short exposure (the bird blips showing on the picture as dots) or with a long exposure (echoes of the moving bird blips showing as straight or curved lines) (138, 367). Small birds are often near the threshold of detection, which means that during one antenna rotation they may just "paint" (i.e., produce a visible dot on the radar screen), but may not be visible during the next few radar sweeps. Thus, photos with a short exposure time, taken during nights with many small birds aloft, do not give a true picture. A long exposure, on the other hand, often has the shortcoming that the lines produced by the moving echoes overlap, and produce a confusing picture.

Another much-used method is to make "time-lapse movies": one frame is exposed during one complete revolution of the antenna, the next frame covers the next antenna sweep, and so on.

When the antenna rotates at 5 revolutions per minute, each film frame covers a period of 12 seconds. If the radar film is projected at speeds of 16 frames per second, there is a time compression of 12 x 16 = 192 times. This means that aircraft blips move quickly over the projection screen, and bird blips can be differentiated due to differences in speed, direction, range, and echo size and brightness (138, 367). Movie films thus have the advantage of giving a better impression of the bird movement and allowing a quick estimate of the general echo density, but they generally do not allow the counting of individual echoes, and certainly do not during heavy passerine migration.

CALIBRATION OF RADARS FOR BIRDS

With so many hurdles to surmount it is obvious that there have been few quantitative bird studies using ATC surveillance or similar radar equipment. Few scientists have been able to calibrate their radar for birds, that is, compare the density of bird echoes on the PPI screen with actual bird counts obtained in some other way. In Massachusetts, US, for instance, Nisbet compared moonwatch observations (Chapter 1) with PPI density. The quantification of the bird information involved complicated procedures to determine the rate of thinning of the echoes with range. He concluded that when migration density was low, estimates of bird numbers would have a standard error of about 25 per cent; when density was high, estimates would be less accurate; and that very high densities could not be measured at all, although they could be accurately identified as such (313). Furthermore, the results of this study could not be applied to other radar stations, nor to other species of birds.

Another study to quantify bird data from a PPI screen was carried out by Gauthreaux with a WSR-57 weather surveillance radar in Louisiana, US. He counted birds at night by moonwatching or with the ceilometer technique (described in Chapter 1), and during the day with a vertically mounted telescope. Fig. 6-3 shows the scale used to evaluate the density of bird echoes on the radar screen. The separate scales reflect the difference between the types of bird echoes that most frequently appeared on the radar screen. At night, both types were present, but the fine echoes greatly outnumbered the dot echoes. The fine echoes were often so dense that they saturated the PPI within the 20 nautical mile range marker, and the time exposures of such patterns gave only limited information on movement and direction. Moonwatching and ceilo-

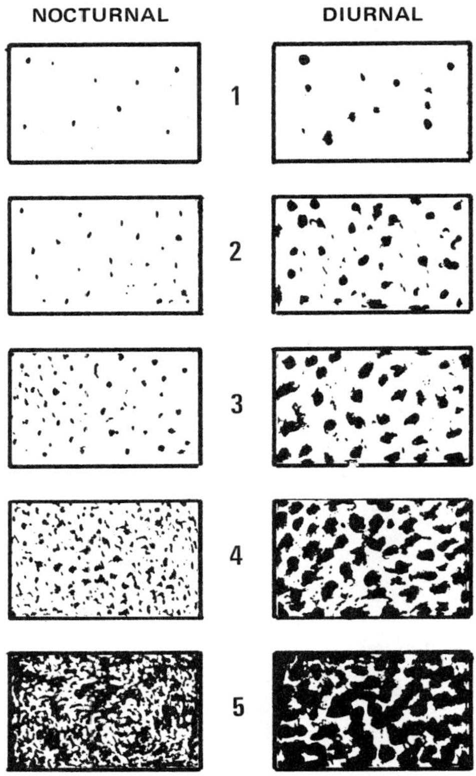

Fig. 6-3: Density patterns of bird echoes on the screen of a WSR-57 weather radar, using a scale of 1 to 5. (left) Nighttime migration. (right) Daytime migration (after 160).

meter observations indicated that passerine birds flying individually in the night sky produced the fine echoes, whereas tight flocks of waterfowl and shore birds caused the scattered dot echoes. Once the radar screen was saturated with echoes, additional echoes could not be displayed even though bird densities increased. To overcome this problem the radar's attenuator (a device to reduce the radar's sensitivity) was used and a sample area on the screen was selected. When the density of bird echoes in the sample area exceeded the base pattern 4 (see Fig. 6-3), Gauthreaux attenuated the signal until this base pattern was reached again, and he recorded the amount of attenuation. He calibrated his radar by plotting the amount of attenuation against the bird density expressed as the Migration Traffic Rate (the number of birds per mile of front per hour; see Chapter 1). For any night with heavy migration, this calibration chart gives the Migration Traffic Rate in accordance with the amount of attenuation used to reach the base pattern (see Fig. 6-4; 160).

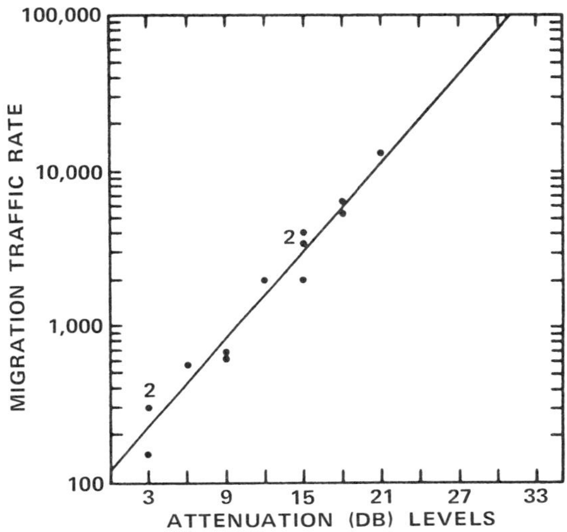

Fig. 6-4: Migration Traffic Rate plotted against the attenuation of the radar signal. Nighttime density pattern 4 was used as the reference base (after 160).

In Europe, progress in quantifying radar bird information has also been made. When it became increasingly obvious that the 0-8 density scale had too many shortcomings to be used for an international network of bird warnings and forecasts (219, 220, 275), the Radar Working Group of the Bird Strike Committee Europe decided in 1972 that the member countries should sponsor studies to arrive at a scale for bird densities that could be applied to different radars used in different geographical areas with various bird groups. Members were urged to concentrate their efforts on electronic techniques for counting birds.

RESEARCH ON "BIRD RADARS" FOR USE IN ATC OPERATIONS

As radar is a useful tool to detect birds in the air, and as airborne birds are a proven hazard to flight safety, it seems logical that special "bird radars" should be developed to give the maximum amount of bird information useful for ATC operations. However, development and production of a new specialized radar is costly. Therefore, most research and development of bird radars has been directed towards improving or refining the bird detecting capabilities of existing radars, especially those already installed at many airports and used for air traffic and weather surveillance. Airborne radars, as installed in the nose of many aircraft, can also detect birds (e.g., 169), but their range is too short and their beam too wide to be useful for bird avoidance purposes.

As indicated earlier, it is difficult to assign definite values to radar bird data as shown on PPI scopes. A different approach is to assess bird density electronically by counting the original, returned signals (the "echoes" received from birds and bird flocks) rather than the processed signals (i.e., the dots or blips on the radar scope). This procedure shortcuts the problem of obtaining a permanent record of the radar screen by filming or photography and largely avoids the problem of visual radar presentation. Furthermore, by using a separate receiver channel (or by intercepting the signals before they are processed by the normal receiver) the problem of most of the radar "fixes" is eliminated.

Work along these lines has been proposed in Holland, discussed in England, and done in Denmark. The Dutch proposal was to divide the space covered by the radar into sectors and to count electronically the total number of returned signals for each sector. An echo is counted when in a resolution cell of the radar the signal exceeds a predetermined level (i.e., how many birds or how big a bird is considered a serious threat). Many of the counted echoes will be due to ground clutter (i.e., returns from stationary targets). To eliminate these echoes, sector counts have to be made at a time when no birds are in the air. By subtracting the latter total from the former, one obtains the total number of bird echoes. Next, a correction factor is introduced to make the counted number of echoes independent of distance, and finally bird densities for each sector (or an average for the whole area) are determined electronically (396). Due to lack of funds, the proposed equipment (including some further refinements, 397) was not developed.

A method proposed in the UK was based on the use of a "Bird Echo Intensity Measuring Receiver," in which an additional receiver channel is used. This system can measure a range of echo intensities of up to 90 dB, whereas the "PPI-photographic" method is confined to a range of 10-20 dB. Fig. 6-5 illustrates how the receiver operates. A range and bearing "gate" (R) is superimposed on the bird movement shown on the PPI. The bird movement,

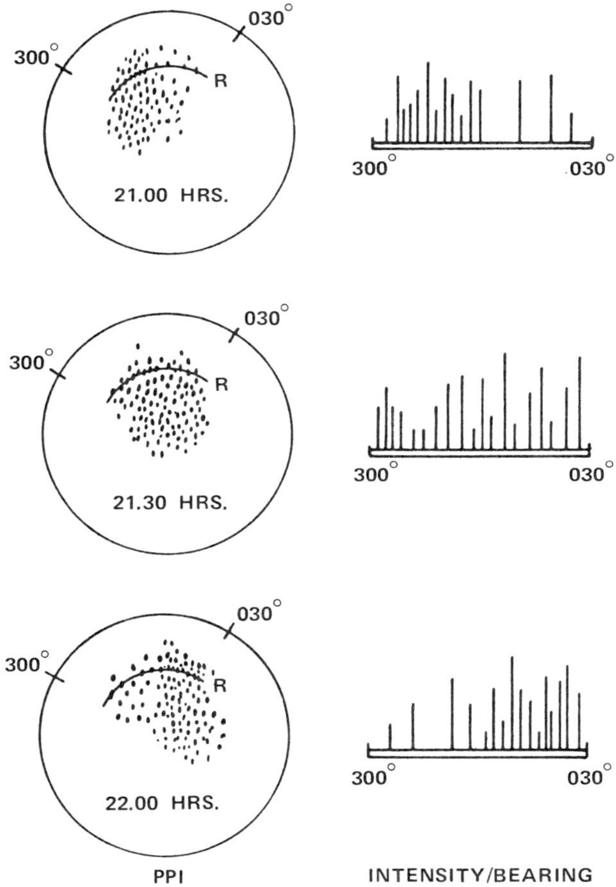

Fig. 6-5: PPI gate indicator and intensity/bearing diagrams, explained in the text on pages 180 to 182 (after 216).

heading north in this case, moves continuously through the gate. Intensities of echoes passing through the gate are recorded at their correct bearings shown on the right-hand diagrams. Echo intensities are recorded as the antenna scans through the gate position. The intensity pattern changes with the passage of time. It is possible to leave the gate position fixed, once it has been positioned on the centre of the main bird movement. The gate is positioned clear of ground and weather clutter, to avoid collecting unwanted data (216).

The intensity and position data are stored on magnetic tape. A computer can be programmed to print-out in chart format the intensities of targets in the gate. If echo intensities are reaching a level that is considered dangerous for flying, the computer can also be programmed to put a warning on the print-out. Another advantage claimed is that the Echo Intensity Measuring Receiver can be easily calibrated before each operation (216). The proposal has not yet been tested.

In Denmark, electronic counting techniques were tested in 1972 (108), improved in 1973 (109), and in operational use during the spring 1974 migration (237). Three "radar windows" were chosen over the sea (i.e., sectors "gated" in range and azimuth). All echoes above a set minimum intensity threshold inside these windows were counted electronically. A log receiver with a wide dynamic range was used. The research and development work was done at Air Station Stensved on behalf of the Danish Bird Hazard Committee. Electronic counting was more reliable, more accurate, faster and cheaper than the photographic system using the 0-8 density scale. Equipment costs (with the exception of the counter) were less than $100.

The term "electronic counting" of birds is somewhat misleading, because the radar merely registers those pulse volumes that appear to present enough echoing area to send back a detectable signal. The echoing area may represent one or more birds or bird flocks. Nevertheless, the electronic counting method is a great improvement over the 0-8 density scale.

In Canada there are two bird migration problems: nighttime migration consisting mainly of small passerine birds, and flights of waterfowl that migrate in large flocks both day and night. To deal with the first problem, plans are underway to count birds electronically using the Danish system. Once the bird density is known, actual bird strike probabilities can be read from specially prepared tables (221). This electronic method can also be used for counting flocks of migrating waterfowl, as was done in spring 1974 at Winnipeg Airport during the annual Snow Goose migration (223).

In areas where there is no surveillance radar, or where the existing ATC radar does not give adequate coverage (because of ground clutter, etc.), a cheap pencil-beam vertical-looking radar can be used to provide height and density of the nighttime migrants.

The bird information obtained from radar using electronic counting techniques is used, not to guide airplanes around airborne bird hazards, but to warn pilots or to divert them to other airfields. Obviously, when a "blanket" of migrating small birds covers a large area, it is impossible to detour aircraft around the hazard. Even when only well-spaced flocks of geese are in the air, all individually painting on the radar screen, it appears difficult to guide incoming and outgoing airplanes around the flocks because of the limited manoeuvrability of big airliners and the presence of other air traffic. The need for close co-operation between ATC, pilots, radar engineers, and biologists in developing as highly specialized a tool as a bird radar is obvious.

CONCLUSIONS

Of the various methods described for reducing the bird hazard en route, the most practical results can be expected to come from bird radars (i.e., radars specially designed, modified, or adapted to "count" birds) in combination with a small calculator, or charts, to convert bird density data into bird strike risk. Such radars have already shown their usefulness in military air operations, and modifications of existing radars need not be costly. Further research and development work in this area is likely to be most fruitful.

There remains the problem of arriving at a set of guidelines on how to deal with various levels of bird collision probability. In addition, ATC personnel and pilots should be instructed during their training about birds, bird movements, and radar bird detection to become familiar with the problem.

Gulls cannot be blamed for the presence of open garbage dumps near airports.

Chapter seven: Bird problem or people problem?

ORGANIZATIONS WORKING ON THE BIRD STRIKE HAZARD

Most countries operating an airline or an air force have a bird hazard problem. However, aircraft accidents involving birds often go unreported. Yet, unless strikes are drawn to the attention of those responsible for providing financial support for preventive or corrective programs, little action will be taken. It is therefore essential to obtain complete reports of any bird strikes on both domestic and visiting aircraft, so that the responsible agency will be fully aware of any hazard involved.

A country can learn whether it has a bird hazard problem by having its national authorities insist that bird strike reporting forms are completed and submitted to the responsible agency. Reporting could be improved in civil aviation by making it mandatory. This will also enable the countries concerned to meet their obligation to submit strike reports to ICAO (see Chapter 2).

Responsibility for the bird hazard problem usually lies with a government agency concerned with civil or military aviation. This varies considerably from country to country, and in many cases the legal responsibilities are not well defined.

In Canada, the federal Ministry of Transport operates most of the large airports. It is responsible for airworthiness, flight safety, and control of bird hazards at airports.

In the US, most airports are operated by city or private organizations; only a few airports are operated by the federal government. This situation has made the problem more difficult, administratively. Recent legislation has made the US Federal Aviation Administration responsible for certifying that all airports are safe, as regards the bird hazard.

Few countries have within one organization all of the know-how and experience required to solve a problem that involves such varied specialities. A procedure that has been shown in practice to be effective is to appoint a National Committee or Working Group.

A NATIONAL COMMITTEE

The problems involved in dealing with bird hazards are complex. Consequently, the members of a national committee should be appointed with the varied nature of the work in mind. They should represent all groups that are directly involved in or concerned with bird strikes, and include scientists experienced in the fields of aircraft engineering and biology. It is important that non-government representatives be included. A main consideration should be to select each representative for his ability to contribute to the committee's work, and not merely to represent an agency or organization. Membership on the committee may be named from the following:

— civil and military aviation (flight safety, airport construction, and airport management divisions)
— major airlines
— the national aircraft industry
— air traffic controllers (possibly from the controllers' association)
— airline pilots (possibly from the pilots' association)
— ornithologists and ecologists from universities and government.

The terms of reference of the National Committee should be sufficiently clear to guide its general direction, but they should not restrict its effectiveness.

Canada, the US, and many European countries now have national committees, which approach their bird hazard problems with varying degrees of determination and success. A list of these committees is given in Appendix 7-1. For countries without a national committee, the names of people active in the field are given in this appendix.

WORK OF A NATIONAL COMMITTEE

A committee can make many contributions towards reducing the bird hazard. The first goal of a new national committee is to identify the problems and assess the situation as it applies within the country concerned. The availability of bird strike statistics will be essential to judge the seriousness of the problem. If bird strikes are not yet reported routinely, a program should be begun to ensure that all bird strike incidents are systematically recorded and analyzed. A detailed discussion of strike reporting is given in Chapter 2.

Based on such information as it may have, the committee can arrange a priority list of problems needing study, and can take steps to obtain financial support for this work.

To solve certain bird problems, the committee can carry out or contract for the necessary research involved, or it can try to apply the results of studies done elsewhere to the local situation.

The committee should make sure that the information obtained through its activities and sponsorship is published and distributed. It could also arrange for manuals or books to be prepared that include up-to-date information, such as have been produced by Britain, West Germany, the USSR, and ICAO (412, 206, 233, and 224, respectively).

In order to introduce new methods and ensure their use, the committee should arrange instructional courses in bird strike prevention to be given to personnel who will be involved, particularly airport staff. This has been done in Canada, the US, and France (338, 161, 432). The committee can also give advice on how to set up an airport bird control team. Another task is to determine the operational effectiveness of the newly introduced methods, and to recommend improvements when needed and feasible.

A continuing public information program should also be maintained under the committee's guidance, as well as publicity directed at aviation personnel, to keep them alerted to the hazard. Such a program can make use of articles, lectures, posters, cartoons, films, etc.

A new airport site should not be selected until a study has been made of the likelihood that birds will be a problem. The probable hazard at alternative suggested sites can be assessed by the committee. Similarly, when extension of existing airport facilities is being considered, the bird hazard should be assessed before a decision to proceed is reached.

INTERNATIONAL COMMITTEES

Some countries have been engaged in work on bird hazard control for a number of years. It is mutually advantageous to national committees and to individuals to keep in touch with other groups. This can be of benefit in exchanging information on work done or in progress, thus helping to avoid duplication.

Many problems are common to various countries, especially in the same geographical areas. Consequently, there are benefits in setting up an International Committee for a given area. An outstanding example of such an international committee is that of Bird Strike Committee Europe (BSCE). This committee was formed in 1966 and has been active in many areas. Membership is open to Western European countries. The US and Canada, as users of European airspace, have been welcomed to meetings. In order to deal more effectively with certain aspects of the problem, a few working groups were established. Appendix 7-2 gives the terms of reference and the names and addresses of the chairmen of the main committee and its working groups.

In North America, the US and Canada have a loosely knit association, and free exchange of information takes place.

INTERNATIONAL CIVIL AVIATION ORGANIZATION (ICAO)

The ICAO was set up by the United Nations and deals with some of the problems relating to civil aviation. Over 100 countries are member states, and membership is open to any country. The ICAO Air Navigation Commission, under which the Airworthiness Committee functions, is responsible for problems relating to the bird hazard.

Each member country is represented on ICAO by its State Representative, and bird strike reports should be submitted to him to meet national obligations. It is to the advantage of those interested in the bird hazard problem (or any other aviation problem) to make contact with their own State Representative. In this way they will be able to obtain working papers, reports, and recommendations from ICAO. They will also be able to influence the views of their State Representative. In some countries those working on the bird problem are unaware of how ICAO functions, and as a result they may not receive much useful information.

The Headquarters of ICAO is located in Montreal, Canada, and many of the meetings are held there, but Regional Meetings of ICAO are also held in various parts of the world so that problems that are largely regional can be discussed.

WORLD CONFERENCES

The first international conference on bird hazards to aircraft was held at Nice, France, in 1963. Its proceedings were published (85). A second meeting was held in Frankfurt, Germany, in 1966, which resulted in the formation of the BSCE. Canada held a World Conference on Bird Hazards to Aircraft in Kingston, Ontario, in 1969, and its proceedings are also available (92).

The advantages of such meetings include valuable publicity, exchange of information, and the publication of proceedings to make this information permanently available.

MEASUREMENT CONVERSION TABLE

ENGLISH SYSTEM				METRIC SYSTEM	
Linear					
1 inch			=	2.54	cm
1 foot	=	12 inches	=	30.48	cm
1 yard	=	3 feet	=	91.44	cm
1 mile	=	1,760 yards	=	1.609	km
Surface					
1 acre			=	0.4047	hectare
1 sq. mile	=	640 acres	=	2.590	km^2
Weight					
1 ounce			=	28.35	gm
1 pound (lb)	=	16 ounces	=	0.4536	kg
Temperature					
χ °F			=	$(\chi - 32) \times \frac{5}{9}$ °C	

Appendices to Chapters 1 through 7

Appendix 1-1: Weights and flock densities for some bird species (Ref. 424)

SPECIES	WEIGHT (LB.)	FLOCK DENSITY (LB./1,000 CU.FT.)	
		AVERAGE	MAXIMUM
American White Pelican	15.0	12	31
Gannet	5.7	7	15
Mute Swan	35.3	66	108
Whistling Swan	16.0	22	31
Brant Goose	3.1	70	150
Grey Lag Goose	7.7	13	29
Snow Goose	6.0	55	132
Mallard	2.4	40	126
Black Kite	1.5	5	35
Crane	12.1	12	39
Sandhill Crane	12.0	15	28
European Oystercatcher	1.2	59	238
Eurasian Golden Plover	0.46	35	67
Spotted Redshank	0.33	63	86
Knot	0.24	145	226
Sanderling	0.13	41	79
Rufous-necked Sandpiper	0.033	39	111
Black-headed Gull	0.65	16	45
Silver Gull	0.72	24	52
Rock Dove (pigeon)	0.85	58	230
Wood Pigeon	1.05	132	315
Common Swift	0.087	2	3
Cliff Swallow	0.046	14	18
Common Starling	0.17	52	153

Appendix 1-2: Heights of migrating birds obtained from radar studies. All heights are above ground level.

LOCATION	MAIN HEIGHT INFORMATION	REFERENCE
EUROPE		
England	morning and evening migration common at 2,000 to 3,000 ft., frequent up to 5,000 ft. and very infrequent above 10,000 ft. Peaks at 13,000, 14,000 and 16,000 ft.	(186)
England	passerine migration mainly below 5,000 ft., occasionally up to 14,000 ft.	(251)
Finland	Common Scoters and Long-tailed Ducks at 600-900 ft. when over the sea, but at 1,500 to 6,600 ft. when over land.	(37)
Southern North Sea	medium altitude 2,000 ft. by day and 1,800 by night.	(140)
Switzerland	most daytime migration under 3,000 ft.	(162)
Switzerland	50% of nighttime migration below 2,200 ft. and 90% below 6,600 ft. Highest birds at 13,200 ft. in spring and 14,800 ft. in fall.	(80)
NORTH AMERICA		
Alberta, Canada	nighttime migration for 50% below 2,500 ft., 90% below 5,000 ft., and 99% below 10,000 ft.	(53)
Coast of Gulf of Mexico, US	nighttime passerine migration for 75% below 3,000 ft. with 90% below 5,000 ft.	(1)
Eastern Canada	nighttime migration of passerines and waterfowl mainly below 4,000 ft., occasionally up to 7,000 ft. Shorebirds migration much higher; not infrequently up to 15,000 ft. and occasionally at 18-21,000 ft.	(332)
Illinois, US	nighttime migration up to 6,500 ft. in spring and up to 7,000 ft. in autumn.	(35)
Massachusetts, US	nighttime migration usually at 1,500 to 2,500 ft., 90% below 5,000 ft. and 99% below 10,000 ft.	(314)

Appendix 1-3: Groundspeeds of some bird species (Ref. 289)

SPECIES	LOCATION	GROUNDSPEED (MPH)	REMARKS
Carrion Crow	Silesia	29	
Carrion Crow	UK	20-32	
Carrion Crow	Germany	31.2	migrating
Common Starling	UK	23-32	cruising
Common Starling	Palestine	45-48.5	flying to roost
Common Starling	Baluchistan	43-49	migrating
Chaffinch	Germany	32.8	migrating
Chaffinch	UK	21, 22, 22	normal flight
Blackbird	UK	22-33	
Blackbird	UK	29, 30	alarmed
Bluebird	US	13-26	
Chipping Sparrow	New Hampshire	15, 15, 20	
Skylark	UK	22-28	
Skylark	Arabia	33-37	migrating
Great Blue Heron	US	19-28	
Peregrine Falcon	UK	62	level cruising
Peregrine Falcon	Germany	37	migrating
Bewick's Swan	UK	39-42	normal flight
Mute Swan	UK	31, 37, 38, 40	normal flight
"Geese"	US	44.3	migrating
Snow Goose	California	50	
Mallard	US	46-70	
Pintail	Sudan	37-51	
Long-tailed Duck	Canada	54-72	
Wood Pigeon	Germany	33-54	
Wood Pigeon	UK	41-43	normal flight
Wood Pigeon	UK	50, 51	hard-pressed

Appendix 2-1: **Number of bird strikes in civil aviation for some countries (compiled by M.S. Kuhring from reports to ICAO and correspondence with various countries).**

	1968	1969	1970	1971	1972	1973
Australia	160	198	173	116	99	—
Belgium	7	—	—	10	—	—
Canada	232	281	271	278	258	278
Chad	—	—	1	—	—	1
Czechoslovakia	—	—	—	5	—	—
Central Africa Republic	1	1	—	—	—	—
Dakar	14	—	—	—	—	—
Denmark	19	32	35	48	59	48
East Africa (Kenya, Uganda, Tanzania)	35	35	—	6	11	—
Finland	8	8	—	—	—	—
France	52	68	75	84	85	56
Ghana	—	—	3	3	5	—
Hong Kong	6	12	20	1	—	—
India	25	35	39	36	49	48
Ireland	10	10	29	9	9	—
Israel	15	16	11	—	—	—
Italy	—	—	—	4	—	—
Ivory Coast	1	—	—	—	—	—
Japan[a]	20	41	44	37	55	—
Malawi	6	10	—	—	—	—
Malaysia	—	—	—	3	—	—
Mexico	—	—	2	—	—	—
Netherlands	194	181	148	137	176	151
New Zealand	98	114	117	163	107	153
Niger	3	2	1	—	—	—
Nigeria	—	—	—	1	—	—
Norway[b]	12	—	50	25	33	—
Pakistan	6	3	7	5	7	3
Portugal	—	—	41	—	—	—
South Africa	26	30	—	73	49	43
Spain	2	—	—	—	—	—
Sweden	36	36	30	55	63	—
Switzerland[c]	—	—	35	96	90	—
United Kingdom	275	246	307	336	361	—
USSR[d]	—	221	153	108	113	152
United States	504	—	—	286	246	—
West Germany	—	—	221	312	312	—

a) Includes both civil and military
b) Norwegian-registered aircraft only
c) 1972 data estimated from May/Dec. reports
d) Estimated as about 1,500 for whole USSR
— Unknown or not given

Appendix 2-2: Number of bird strikes on military aircraft for some countries (compiled by M.S. Kuhring).

	1966	1967	1968	1969	1970	1971	1972	1973
Belgium	24	21	48	40	33	44	49	65
Canada	198	254	197	228	167	184	178	180
Denmark	19	—	49	54	41	46	42	52
France	—	81	94	86	58	78	127	141
India	131	129	94	113	124	105	111	67
Italy	—	—	—	—	—	—	—	1
Netherlands	81	108	98	79	86	96	114	94
Norway	14	38	36	27	26	36	34	42
United Kingdom	—	—	—	—	183	210	214	—
United States	320	379	363	338	360	383	351	327
West Germany	—	—	—	—	—	259	—	—

— Unknown or not given

Appendix 2-3: Bird strike data for Air Canada, 1959-73 (after Ref. 47)

	1959/60	1961	1962	1963	1964	1965	1966	1967	1968	1969	1970	1971	1972	1973
Strikes	35/67	99	138	170	164	125	147	187	183	214	208	188	173	185
Near-misses	0	0	0	2	4	3	9	2	4	7	8	14	37	21
Total incidents	35/67	99	138	172	168	128	156	189	187	221	216	202	216	206
Incidents in Canada	92	95	120	148	143	111	139	168	167	204	199	187	196	199
Incidents outside Canada	10	4	18	24	25	17	17	21	20	17	17	15	20	7
Strikes/10,000 Take-offs and landings	0	0	0	6.18	5.98	4.18	4.35	4.83	4.56	5.64	5.26	5.05	4.84	4.63
Strikes/10,000 hr.	2.18/ 3.94	5.59	8.46	10.9	10.38	7.06	7.39	7.57	6.94	8.75	7.97	7.48	7.06	6.78
Strike costs (x $1000)														
In Canada	475,	330,	40,	20,	195,	83,	40,	75,	187,	10,	235,	8,	153,	24,
Outside Canada	25,	20,	140,	140,	8,	5,	31,	0,	0,	40,	0,	190,	1,	0,
Total	500,	350,	180,	160,	203,	88,	71,	75,	187,	50,	235,	198,	154,	24,

Appendix 2-4: Bird species involved in collisions with aircraft in Canada to 1973 (Ref. 371)

Common Loon
Western Grebe
Great Blue Heron
Canada Goose
Mallard
Pintail
Green-winged Teal
Blue-winged Teal
American Widgeon
Red-shouldered Hawk
Bald Eagle
American Kestrel
Gray Partridge
Ring-necked Pheasant
Sandhill Crane
Killdeer
Golden Plover
Black-bellied Plover
Upland Sandpiper
Willet
Dunlin
Baird's Sandpiper
Western Sandpiper
Great Black-backed Gull
Glaucous-winged Gull
Herring Gull
Ring-billed Gull
Franklin's Gull
Rock Dove
Mourning Dove
Great Horned Owl
Snowy Owl
Short-eared Owl
Common Nighthawk
Black Swift
Barn Swallow
Horned Lark
Yellow-shafted Flicker
Blue Jay
Common Crow
Long-billed Marsh Wren
American Robin
Ruby-crowned Kinglet
Audubon's Warbler
Blackpoll Warbler
Western Meadowlark
Red-winged Blackbird
Brewer's Blackbird
Brown-headed Cowbird
Evening Grosbeak
Pine Siskin
American Goldfinch
Purple Finch
Savannah Sparrow
Slate-colored Junco
Harris's Sparrow
Swamp Sparrow
Chestnut-collared Longspur
Lapland Longspur
Snow Bunting

Appendix 2-5: Birds (by group) involved in collisions with aircraft, as percentage of frequency

Country	Holland		United Kingdom	
	RNethAF		RAF	Civil
Period	1956-65	'66—'69	'64—'68	'66—'71
Reference	(50)	(125)	(194)	(406)
Gulls	29.8	30	53.2	64.2
Waders/shorebirds	20.6	28	12.1	15.8
Pigeons	10.1	11	10.7	3.3
Waterfowl	4.3	0	0	0.5
Corvids	4.3	0	7.7	2.7
Raptors	6.6	8	0	2.8
Vultures	0	0	0	0.9
Owls	1.0	0	0	0.5
Pheasants	0.5	0	0	0
Partridges	2.8	0	0	0.4
Starlings	4.8	0	9.8	1.9
Thrushes	1.4	0	0	0.8
Swifts/swallows	9.1	15	7.4	4.9
Cranes/herons	0	0	0	0.1
"Small birds"	4.8	8	0	0.9
Terns	1.0	0	0	0.4

*Includes "small birds"
—Unknown or not given

Appendix 2-5 (cont'd.)

Germany		Sweden	USSR	Canada		Australia	India
GAF '67—'68 (81)	DLH — (81)	RSAF '67—'72 (3)	Civil — (234)	CAF '69—'71 (103)	CP Air '60—'69 (81)	— '58—'65 (423)	Civil '60—'63 (90)
41.8	28.6	37.8	36.3	42.7	36.0	19	0
7.9	11.4	2.7	4.1	7.2	4.5	17	0
8.5	2.9	0	6.7	0.6	2.3	5	7.1
1.8	0	1.8	11.4	6.7	18.2	0	0
4.2	5.7	3.1	6.7	2.8	2.3	0	0
19.5	14.3	2.7	11.3	7.2	0	41	7.1
0	0	0	0	0	0	0	85.7
0.6	8.5	0	4.0	0.6	4.5	0	0
0.6	2.9	0	0	0	0	0	0
1.2	0	0	0	1.7	0	0	0
6.0	11.4	3.3	7.4	3.4	7.0	0	0
		0	0.7	1.1		0	0
7.9	14.3*	15.1	4.7	5.0	7.0	0	0
0	0	0	0.7	0	4.5	0	0
0	0	33.6	4.7	22.2	13.7	18	0
0	0	0	0	0	0	0	0

Appendix 5-1: Planning guidelines for land use outside airports in Canada (Ref. 99)

"**Bird Hazards**

"In the vicinity of airports both individual large birds and flocks of small birds are a potential hazard to the safety of aircraft. At existing airports, this hazard can be minimized by using lands surrounding the airport for purposes that will not attract concentrations of birds to the area.

"At proposed new airports, bird hazards can be minimized by careful selection of the site to avoid established bird migration routes and areas of naturally attractive bird habitat, and by using the lands surrounding the airport for purposes that will not attract concentrations of birds to the area.

"The attached Table I has been prepared to indicate generally those land uses that are either compatible or incompatible with the objective of minimizing bird hazards in the vicinity of airports. (See also Figure 1)

"It should be recognized, however, that these are general guidelines only and that the compatibility or incompatibility of use of land depends heavily on management of the area to prevent attracting birds. For example, one piggery in the Maritimes is totally enclosed and is, therefore, a compatible use because it neither attracts nor supports birds.

"**NOTE:** When considering a specific land use, the recommendations of all sections of this part must be examined and correlated as necessary."

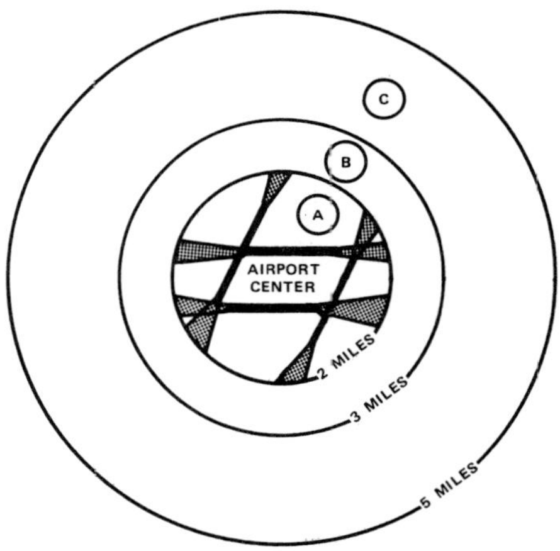

TABLE 1

"Land uses around airports compatible with minimum bird hazards to aircraft

"USES	SOME COMPATIBLE USES (NOTE 2)	INCOMPATIBLE USES	AREAS AFFECTED (NOTE 1)
"Natural (Note 4)	Coniferous forest reserves	Forest reserves	B
		Fish reserves	B
		Bird sanctuaries	C
		Swamps	C
		Flood and flood control areas	C
		Game preserves	C
"Agricultural	Landscape nurseries	Sod farming	B
	Christmas tree farming	Seed farming	B
	Stock farming) (Note 5)	Crop farming	B
	Dairy farming) (Note 5)	Piggeries	C
		Fruit tree farming	C
		Stock feedlots	C
		Stockyards	C
"Recreational (Note 6)	Golf courses	Picnic and camping areas	A
	Parks	Fairgrounds	C
	Playgrounds	Racetracks	B
	Athletic fields	Outdoor theaters	C
	Riding academies & trails		
	Tennis, Lawn Bowling courts		
"Municipal Utilities	Water treatment	Water storage (open)	C
	Sewage treatment plants	Raw sewage outfalls	C
	Gas and oil facilities	Power lines	Ai
	Coal yards	Electrical plants	Ai
	Mobile equipment yards	Garbage dumps	C
		Sewage lagoons	C
		Sanitary landfills	C
"Industrial (Note 6)	Research laboratories	Food processing plants where edible wastes are accessible to birds	B
	Aircraft factories		
	Non-Aircraft factories		
	Assembly plants		
	Transportation terminals		
	Air Freight terminals		

TABLE 1 (CONT)

"Residential and Institutional (Note 6)	Schools Hospitals Churches Family dwellings Residential — 1 acre and above		
"Commercial (Note 6)	Aerial Surveying Offices and Plant Aircraft Sales Aviation Schools Aviation Repairs Aircargo Facilities Hotels and Motels Airport Restaurants Retail Businesses Shopping Centers Banks Service Stations Auto Storage Parking Lots Office Buildings Theatres Auditoriums Stadiums Public Buildings Taxi Terminals Bus Terminals Truck Rentals Memorial Parks	Drive-in burger and food service stands where food is consumed in automobiles and remains are disposed of in open garbage containers.	B
		Flat roofed buildings with inadequate drainage or those designed to hold water on roof. (Acceptable if water not accessible to birds).	A
"Transportation Uses (Note 6)	Highways and Expressways Railroads and Railway Yards Port Facilities		

"NOTE

(1) Area A — land area within a radius of 2 miles of center of airport.
Area B — land area within a radius of 3 miles of center of airport.
Area C — land area within a radius of 5 miles of center of airport.

Ai — that Sector of area A that is delineated as an approach or departure zone by the Flightway obstruction requirements.

(2) All compatible uses, with respect to bird hazards alone, are suitable anywhere within the 5-mile area.

(3) For major international airports, radius of circles delineating Areas A, B and C should be increased by 1 mile.

(4) Provided management does not create or maintain bird populations which create hazards to aircraft safety.

(5) Provided feed is not accessible to birds and that precautions are taken to ensure that disposal of excrement from livestock does not attract birds.

(6) Provided areas are kept clean and free of box lunch remains, restaurant garbage, and other wastes edible by birds."

Appendix 7-1: National groups and individuals concerned with or working on bird strike problems in 1974 (compiled with assistance of M.S. Kuhring)

A. Countries with a National Committee

BELGIUM
Belgium Bird Strike Committee
Chairman — Major W. Lemaire, Meteorological Wing of the Air Force, Lange Eikstraat 86, 1970 Wezembeek.

CANADA
Associate Committee on Bird Hazards to Aircraft, National Research Council, Montreal Road Laboratories, Bldg. M-58, Ottawa
Chairman — Dr. V.E.F. Solman

DENMARK
Bird Strike Committee of the Royal Danish Air Force
Chairman — A.H. Joensen, Game Biology Station, Kalø, 8410 Ronde.

FRANCE
Interministerial Working Group on Bird Problems
Chairman — M. Hoerter, Inspection Générale de l'Aviation Civile, 246 rue Lecourbe, F-75, Paris (15).

HOLLAND
Working Group to Prevent Bird Strikes
Chairman — J.C.N. van Boldrik, Flight and Ground Safety Section, Air Staff, 1ste van den Boschstraat 8, The Hague.

SWITZERLAND
Swiss Bird Strike Committee
General Coordinator — M. Schneiter, Swiss Air Office, Airport Section, Bundeshaus, Inselgasse, 3003 Bern.

UNITED KINGDOM
Bird Impact Research and Development Committee
Chairman — D.A.L. Robson, Ministry of Defence/Procurement Executive (Mech. 2), Room 116, St. Giles Court, St. Giles High Street, London WC 2.

UNITED STATES OF AMERICA
Inter-Agency Bird Hazard Committee
Chairman — J.T. Morse, Safety Standards Branch, Federal Aviation Administration, 800 Independence Ave., S.W., Washington, D.C., 20590.

UNION OF SOVIET SOCIALIST REPUBLICS
Bird Strike Committee of the USSR
Chairman — Dr. V.E. Jacoby, Academy of Sciences, A.N. Svertzov Institute of Evolutionary Morphology and Animal Ecology, Lenin Ave. 33, Moscow W-71.

WEST GERMANY
German Committee for the Prevention of Bird Strikes in Air Traffic
Chairman — Dr. W. Keil, Steinauerstrasse 44, Fechenheim, D6000 Frankfurt am Main 61.

B. Countries without a National Committee

AUSTRALIA
Dr. G.F. van Tets, Commonwealth Scientific and Industrial Research Organization, Division of Wildlife Research, P.O. Box 109, Canberra ACT2601.

H.G. Lavery, Animal Health Station, Dept. of Primary Industries, P.O. Box 1085, Townsville, Queensland.

CEYLON
Director of Civil Aviation, P.O. Box 535, Colombo 1.

CZECHOSLOVAKIA
Z. Veverka, Czechoslovak Airlines, Prague 6 — Ruzyne Airport

GHANA
Director, Dept. of Civil Aviation, Kotoka Airport, P.O. Box 87, Accra.

GREECE
Royal Hellenic Air Force, General Air Staff, Flight Safety Directorate, Holargos, Athens.

HONG KONG
J.D. Romer, Pest Control Officer, Urban Services Dept., Central Government Offices (West Wing).

HUNGARY
Director, Hungarian Institute of Ornithology, Kolto Ucta 21, Budapest XII.

INDIA
Director of Air Safety, Office of the Director General of Civil Aviation, East Block No-3, Level No 4, R.K. Puram, New Delhi, 110022.

Director of Flight Safety, Indian Force, Air Headquarters, New Delhi II.

IRELAND
Capt. W.G. Black, Air Safety Officer, Irish International Airlines, Dublin Airport, Dublin.

Dr. M. Carr, Dept. of Transport and Power, Aeronautical Section (Operations), Dublin.

ISRAEL
T. Elsner, Civil Aviation Administration, Operations Section, Lod Airport, P.O. Box 8.

ITALY
Major N. Bonifacio, Italian Air Force, Corso Regina Maria Pia 18/A, 00056 Roma Ostia.
Director, Directorate of Materials, Leonardo da Vinci Airport, Rome.

JAPAN
Dr. N. Kuroda, Yamashina Institute for Ornithology, Nagahisa, Kumoda, 8-20 Nanpeidai, Shibuya.

Wasna Otsuka, Director of Admin. Division, Aerodrome Department Ministry of Transport, Kasumigaseki, Chiyodaku, Tokyo.

LEBANON
Director, Civil Aviation Centre, Beirut International Airport, Beirut.

MEXICO
J.M. Bellot, Aeronaves de México, S.A., Central 161, Mexico 9, D.F.
J. Soto-Reyes, Aspa de México, Palomas 110, Mexico 10, D.F.

NEW ZEALAND
T.A. Caithness, N.Z. Wildlife Service, Internal Affairs Dept., Private Bag, Wellington.

Mr. Bull, Animal Ecology Division, Department of Scientific and Industrial Research, Lower Hutt.

NORWAY
A.H. Hestnas, R. Nor. Air Force, Air Base Graatallen.
G. Lid, Zoological Museum, Sargst 1, Oslo 5.
P.W. Pedersen, Directorate of Civil Aviation, Storgt. 10B, Oslo.

PAKISTAN
M.A. Khan, Flight Safety Inspection, Pakistan International Airlines, Karachi Airport.

POLAND
Dr. M. Luniak, Polish Academy of Sciences, Institute of Zoology, Warsaw, Wilcza 64.

PORTUGAL
Director of Technical Services, Directorate of Civil Aviation, Av Liberdad 193, Lisbon.

Portuguese Air Force Headquarters, Rus Rodrigues Sampaio 99, Lisbon.

RHODESIA
Director, Dept. of Civil Aviation, P Bay 7716, Causeway, Salisbury.

SPAIN
Dr. D. Ramos, Iberia Air Lines, Velasquez 130, Madrid 6. Director, Flight Safety Section, Ministry of Aviation, Madrid.

SOUTH AFRICA
Prof. J.M. Winterbottom, University of Cape Town,
Percy Fitzpatrick Institute, Ronde Bosch CP, Cape Town.

Dr. C.G. van Nierkerk, Aeronautics Research Unit, P.O. Box 395, Pretoria.

SWEDEN
N.A. Bergquist, R. Swedish Air Force Headquarters, Flight Safety, Stockholm.

L.A. Turesson, Board of Civil Aviation, Stockholm.

Dr. S. Ulfstrand, Dept. of Zoology, University of Lund, Lund.

Appendix 7-2: Bird Strike Committee Europe and its Working Groups in 1975 (Information Circ. 34/BSCE/10WP/1, 14 Jan. 1975)

Terms of reference

"In agreeing to the establishment of the Bird Strike Committee Europe at a meeting held in 1965, the representatives of the National Bird Strike Committee of some European countries and the experts attending the annual meeting of the Flight Safety (Air Force) Europe agreed that the Committee's terms of Reference should be

a) collect, analyse all datum related to bird activities in Europe;
b) study, prepare and circulate methods to scare birds on and in the vicinity of airports (airfields);
c) explore the feasibility of interpreting bird echoes on radar screen and develop a method of warning crews when hazard is positively identified. Try to develop also avoidance methods that could be initiated and controlled by radar operators;
d) ensure a quick and safe circulation of messages covering bird hazards;
e) develop the associated information to be included in Aeronautical Information Publications;
f) standardize all over Europe methods which have been tested and found acceptable;
g) encourage studies to improve safety in flight, regardless of the field of application, under the unique condition that it could be used by all European countries;
h) make available to all countries the findings and results achieved."

CHAIRMAN IN 1975

V. E. Ferry, Bird Strike Committee Europe, Inspection Générale de l'Aviation Civile, 246 rue Lecourbe, #75732 Paris Cedex 15 (Tel.: 828-34-20).

WORKING GROUPS IN 1975

Bird Movement Working Group

"Study of bird concentration and movements and the drawing up of special maps for the information of aircrews and aviation services."

Chairman: Dr. J. Hild, Geophysical Office, GAF, 5050 Porz-Wahn, Post Box 5000/507, West Germany.

Communications Working Group

"Technical subcommittee dealing with matters associated with the use of radar in the surveillance, identification and assessment of bird movements.

The work embraces the evaluation of the radar properties of birds, of specific radar equipments and techniques, of improvements in design and data handling, and also of 'best' ways of operating radars."

Chairman: E. W. Houghton, Royal Radar Establishment, St. Andrews Road, Great Malvern, Worc., UK.

Aerodrome Working Group

"a) Preparation of general recommendation to minimize bird problems on and around aerodromes.

b) Correlation of bird control research between the countries."

Chairman: Dr. W. Keil, Steinauer Str. 44, D6000 Frankfurt am Main 61, W. Germany.

Analysis Working Group

"To agree on a standardised format for the production of Analyses from the data contained in Bird Strike Report forms."

Chairman: J. Thorpe, Civil Aviation Authority, Redhill, Surrey, UK.

List of Abbreviations

ABDE	Airport Bird Detection Equipment
AFB	Air Force Base
agl	above ground level
asl	above sea level
ATC	Air Traffic Control
BCAR	British Civil Aviation Requirements
BOAC	British Overseas Airline Corporation
CAF	Canadian Armed Forces
CFB	Canadian Forces Base
CWS	Canadian Wildlife Service
dB	decibel
DLH	Deutsche Luft Hansa
FAA	Federal Aviation Administration (USA)
FAR	Federal Aviation Requirements (USA)
FOD	Foreign Object Damage
GAF	German Air Force (West Germany)
GCA	ground control approach
Hz	Hertz (cycles per second)
ICAO	International Civil Aviation Organization
IR	infrared
KLM	Koninklyke Luchtvaart Maatschappy (Royal Dutch Airlines)
LLLTV	low light level television
MOT	Ministry of Transport
mph	miles per hour
MTR	Migration Traffic Rate
NAE	National Aeronautical Establishment
PPI	Plan Position Indicator
PVB	polyvinyl butyral
RAF	Royal Air Force (United Kingdom)
RNethAF	Royal Netherlands Air Force
RNorAF	Royal Norwegian Air Force
rpm	revolutions per minute
RSwedAF	Royal Swedish Air Force
SpanAF	Spanish Air Force
TAS	true air speed
UK	United Kingdom
US Navy	United States Navy
USAF	United States Air Force
USSR	Union of Soviet Socialist Republics
VFR	Visual Flight Rules

References

ACBHA — Associate Committee on Bird Hazards to Aircraft (National Research Council of Canada, Ottawa, Ontario)

BSCE — Bird Strike Committee Europe (see App. 7-2)

1 Able, K.P. 1970. "A radar study of the altitude of nocturnal passerine migration." *Bird Banding* 41:282-290.

2 Ahlers, R.H. 1966. "Stability of structure following bird strike." Tech. Rept. ADS-60. Aircraft Development Serv., USA.

3 Alerstam, T. 1973. "Bird strikes in the Royal Swedish Air Force, 1967-1972." Issued by Univ. of Lund, Sweden. n.p. (in Swedish).

4 Alerstam, T. 1973. Personal communication.

5 Alerstam, T. and C.A. Bauer. 1973. "A radar study of the spring migration of the Crane *(Grus grus)* over the southern Baltic area." *Vogelwarte* 27:1-16.

6 Alerstam, T., S.G. Lindgren and S. Ulfstrand. 1973. "Nocturnal passerine migration and cold front passages in autumn—a combined radar and field study." *Ornis Scandinavica* 4:103-111.

7 Alerstam, T. and S. Ulfstrand. 1972. "Radar and field observations of diurnal bird migration in South Sweden, Autumn 1971." *Ornis Scandinavica* 3:99-139.

8 Aldrich, T.W. 1964. "The gooney birds of Midway." *Nat. Geog. Mag.* (6):838-851.

9 Allcock, A.W.R. 1969. *In* 92, pp. 389-390.

10 Allcock, A.W.R. 1969. *In* 92, p. 411.

11 Anonymous. 1964. "Bird hazards in Trans-Canada Airlines operations 1959-63." Attach. to State Letter AN3/32-65/161 (Air C-WP 229).

12 Anonymous. 1964. "Bird costs £1,000,000." *Sunday Express*, 24 May 1964.

13 Anonymous. 1968. "They are always around." *Veilig Vliegen* (4):84-85. (in Dutch).

14 Anonymous. 1968-69. "One way to solve a problem." *U.S. Safety Kit.* Dec/Jan 1968/'69.

15 Anonymous. 1972. "The airport dilemma." *Time*, 13 Nov 1972.

16 Anonymous. 1972. "Midairs with birds." *Approach, Naval Aviation Safety Review*, March 1972. pp. 26-33.

17 Anonymous. 1973. "Operation Lights On." *FAA News*. Oct 1973. pp 3.

18 Anonymous. 1974. "CAF Musketeers 'almost trouble free'." *Can. Aviation*. Jan 1974.

19 Anonymous. 1974. "Bird ingestion, fuel dumping." *Flight Safety Focus*. April 1974. pp 13-15.

20 Armstrong, E.A. 1963. *A study of bird song.* Oxford University Press. London.

21 Atkinson-Willes, G.L. 1971. "Report on the January Census of Palaearctic Wildlife in Europe, South-West Asia, and North Africa, 1967 and 1968." Proc. Int. Reg. Mtg. Cons. Wildlife Resources, Leningrad, USSR, 25-30 Sept 1968: pp. 221-238

22 Austin-Smith, P.J. (undated). "Alternative vegetative ground cover for airports in eastern Canada." Rept. to Can. Wildlife Serv. and Nat. Res. Council. Ottawa. 219 pp.

23 Bakke, T.A. 1972. "Food of the Common Gull, *Larus canus* L., and the Black-headed Gull, *Larus ridibundus* L., at Sola airport, Rogaland County." *Fauna* 25:197-204.

24 Ball. 1973. Personal communication.

25 Beason, R.C. 1972. "Waterfowl migration corridors." Air Force Weapons Laboratory, Air Force Systems Command, Kirtland AFB, New Mexico. Tech. Rept. AFWL-TR-72-166, 6 pp.

26 Beason, R.C. 1973. "The bird-aircraft strike hazard in the Canal Zone." Air Force Weapons Laboratory, Air Force Systems Command, Kirtland AFB, New Mexico. Tech. Rept. AFWL-TR-72-16, 20 pp.

27 Beaumont, P.N. and S. Parker. 1973. "Windshield design concepts." *In* 414, pp. 3-13.

28 Beer, C.G. 1960. "Removal of birds from airfields." UK Ministry of Aviation Report, March 1960.

29 Bellrose, F.C. 1966. "Radar in orientation research." *In Proc. XIV Internat. Ornithol. Congr.*:pp.281-309.

30 Bellrose, F.C. 1967. "Establishing certain parameters of hazards to aircraft by migrating birds in the Mississippi Flyway." Rept. for US Dept. Interior, Sport Fish. Wildlife Div., Wildlife Res. Bur., Washington, D.C. 74 pp.

31 Bellrose, F.C. 1968. "Waterfowl migration corridors east of the Rocky Mountains in the United States." Illinois Nat. Hist. Surv., Biol. Notes 61,24 pp.

32 Bellrose, F.C. 1971. "Migration corridors of waterfowl in the United States." US Federal Aviation Administration. Systems Res. Dev. Service, Rept. FAA-RD-71-71. 18 pp.

33 Bellrose, F.C. 1971. "The distribution of nocturnal migrants in the air space." *Auk* 88:397-424.

34 Bellrose, F.C. 1970. Personal communication.

35 Bellrose, F.C. and R.R. Graber. 1973. "A radar study of the flight directions of nocturnal migrants." *In Proc. XIII Internat. Ornithol. Congr.*:pp.362-389.

36 Bellrose, F.C. and J.G. Sieh. 1960. "Massed waterfowl flights in the Mississippi Flyway, 1956 and 1957." *Wilson Bull.* 72:29-59.

37 Bergman, G. and K.O. Donner. 1964. "An analysis of the spring migration of the Common Scoter and the Long-tailed Duck in southern Finland." *Acta Zool. Fenn.* 105:3-59.

38 Berry, B. 1972. "Reproduction by artificial insemination in captive American Goshawks." *J. Wildlife Mgmt.* 36:1283-1288.

39 Berry, H. 1972. "Pelicans air freight their fish 100 kilometers." *African Wildlife* 26:120-124.

40 Bird Strike Committee Europe. 1973. "Recommendations." *In* Proc. 8th Mtg., BSCE, Paris. May 1973.

41 Bird Strike Committee Europe. 1973. "Statistiques de collisions oiseaux/aéronefs." Rept. to 8th Mtg., BSCE, Paris. May 1973.

42 Bird, W.H. 1962. "Bird Hazards in TCA's (Air Canada) operations." Prelim. Rept. to ACBHA.

43 Bird, W.H. 1963. "Bird hazards in Trans-Canada Air Lines operations 1959-1963." *In* 71, pp. 48-67.

44 Bird, W.H. 1964. Letter to V.E.F. Solman, 24 July 1964.

45 Bird, W.H. 1969. Letter to Secretary, ACBHA, 21 Mar 1969.

46 Bird, W.H. 1972. "See and be seen, be seen, be seen by the birds." *The Grapevine*, Air Canada, May-June 1972.

47 Bird, W.H. 1974. "Summary of bird strike statistics, Air Canada, 1959-1973." Rept. for ACBHA.

48 Bird, W.H. 1974. Personal communication.

49 Blaine, G. 1970. *Falconry*. N. Spearman, London, England. 253 pp.

50 Blokpoel, H. 1966. "Survey of bird strikes in the Royal Netherlands Air Force, 1956-1965." Issued by Flight Safety Section, RNethAF. The Hague. 24 pp. (in Dutch with English summary).
51 Blokpoel, H. 1967. "Biological survey of airbase Lahr, West Germany." Rept. for ACBHA. 3 pp.
52 Blokpoel, H. 1967. "The American White Pelican at Primrose Lake in respect to flight safety. A preliminary report." Rept. to Can. Wildlife Serv. 4 pp.
53 Blokpoel, H. 1969. "A preliminary study of height and density of nocturnal fall migration." *In* 92, pp. 334-348.
54 Blokpoel, H. 1973. "Bird migration forecasts for military air operations." Can. Wildlife Serv. Occ. Paper 16. 17 p.
55 Blokpoel, H. 1974. "Migration of Lesser Snow and Blue Geese in spring across southern Manitoba. Part 1. Distribution, chronology, directions, numbers, heights and speeds. Can. Wildlife Serv. Rept. Ser. No. 28. 29 pp.
56 Blokpoel, H. and J. Burton. (in prep). "Weather and the height of nocturnal migration: a radar study."
57 Blokpoel, H. and P.P. Desfosses. 1970. "Radar observations of local bird movements near Calgary, Alberta." ACBHA. Field Note 53, 18 pp.
58 Blokpoel, H. and M.C. Gauthier. (in prep). "Predictions of the 1974 spring Snow Goose migration at Winnipeg International Airport."
59 Blokpoel, H. and M.C. Gauthier. (at press). "Migration of Lesser Snow and Blue Geese across southern Manitoba in spring. Part 2. Influence of the weather and prediction of major flights." Can. Wildlife Serv. Rept. Ser. No. 32. 30 pp.
60 Blokpoel, H., D.J. Heyland, J. Burton and N. Samson. (at press). "Observations of the fall migration of Greater Snow Geese across southern Quebec." *Can. Field-Nat.* 89:268-277.
61 Bosik, A.J., J.B.R. Heath and M.R. Gleeson. 1974. "Capabilities of the NAE/NRC flight impact simulator facility." Nat. Aeronaut. Estab., Nat. Res. Council Can., Ottawa. Internal Rept. LTR-ST.701. 9 pp.
62 Boudreau, G.W. 1968. "Alarm sounds and responses of birds and their application in controlling problem species." *Living Bird* 7:27-46.
63 Boulter, M.J. and R.J. Sobieralski. 1972. "Preliminary evaluation of the pest bird problem at Wright-Patterson AFB, Ohio." Air Force Weapons Laboratory, Air Force Systems Command, Kirtland AFB. New Mexico. Tech. Note DE-TN-72-022, 10 pp.
64 Boulter, M.J., D.E. Compton and G.E. Meyer. 1973. "Evaluation of bird-aircraft collisions at Vance AFB and Kegelman Auxiliary Field, OK. Air Force Weapons Laboratory, Air Force Systems Command, Kirtland AFB, New Mexico." Tech. Note AFWL-TN-73-013, 13 pp.
65 Bourne, W.R.P. 1969. "Gulls and London's Airports." *BTO News.* 17 April 1969.
66 Bourne, B. and P. Rudge. 1970. "Foulness: airport or nature reserve?" *Animals*, Feb. 1970.
67 Bremond, T.-C., P. Gramet, T. Brough and E.N. Wright. 1968. "A comparison of some broadcasting equipments and recorded distress calls for scaring birds." *J. Applied Ecology* 5:521-529.
68 Brewer, G.I. 1970. "Effect of high-intensity flashing lights on pigeon." MSc Thesis. University of Missouri, Missouri. 38 pp.
69 Bridgeman, C.J. 1962. "Birds nesting in aircraft." *British Birds* 55:461-470.
70 Brough, T. 1967. "Recent developments in bird scaring on airfields." *In* 301, pp. 29-38.

71 Brough, T. 1969. "The dispersal of starlings from woodland roosts and the use of bio-acoustics." *J. Applied Ecology* 6:403-410.

72 Brough, T. 1969. "Damage by gulls." *In* Gulls as pests. A discussion organized by the Seabird Group. *Ibis* 111:445-446.

73 Brough, T. 1970. "Bird-strike hazard at Foulness. Study 5." *In* Papers and Proc. VIII, Section 3, Stage III, Res. Inves., Specially Commissioned Studies, Commission on Third London Airport, London.

74 Brough, T. 1971. "Experimental use of long-grass in the UK." Paper for 6th Mtg., BSCE, Copenhagen. June 1971.

75 Brough, T. 1973. Personal communication.

76 Brown, R.G.B. 1962. "The reactions of gulls (Laridae) to distress calls." *Annales des Epiphyties* 13:153-155.

77 Brown, R.G.B. and R.W. Sugg. 1961. "Experiments on the removal of birds from airfields by distress calls (Part 1)." UK Min. Aviation Rept. Aug 1961.

78 Brown, R.G.B., R.W. Sugg and T. Brough. 1962. "Experiments on the removal of birds from airfields by distress calls (Part 2)." UK Min. Aviation Rept. Nov. 1962.

79 Bruderer, B. 1969. "Zur Registrierung und Interpretation von Echosignaturen an einem 3-cm Zielverfolgungs-radar." *Ornithol. Beobach.* 66:70-88.

80 Bruderer, B. 1971. "Radarbeobachtungen über den Frühlingszug im Schweizerischen Mittelland." *Ornithol. Beobach.* 68:89-158.

81 Bruderer, B. 1972. "Vögel und Luftverkehr." Rept. for Swiss Air Office. 39 pp.

82 Bryant, H.E. 1970. "Final report to the Airworthiness Committee, ICAO, on Bird Hazards to Aircraft"; ICAO Document 8458-AN/881. Para. 6. 2. 3. 4. 9th Mtg., Airworthiness Comm., ICAO, Montreal, Nov-Dec 1970.

83 Bub, H. 1968. *Vogelfang und Vogelberingung.* Inst. f. Vogelforschung, Vogelwarte Helgoland, Wilhelmshaven. 116 pp.

84 Burton, J. 1973. Personal communication.

85 Busnel, R.G. and J. Giban (Eds.). 1963. *Le problème des oiseaux sur les aérodromes.* Institute National de la Recherche Agronomique, Paris. 326 pp.

86 Cade, T.J. 1973. "Captive breeding—the 1973 season." Peregrine Fund, Newsletter #1.

87 Caithness, T.A. 1968. "Poisoning gulls with alpha-chloralose near a New Zealand airfield." *J. Wildlife Mgmt.* 32:279-286.

88 Caithness, T.A. 1969. "Research on bird hazards to aircraft in New Zealand." *In* 92, pp. 93-99.

89 Caithness, T.A., M.J. Williams and R.M. Bull. 1967. "Birds and aircraft: a problem on some New Zealand airfields." *Proc. New Zealand Ecol. Soc.* 14:58-62.

90 Canada. National Research Council. 1964. "Reported bird damage incidents in India over the past four years." ACBHA, Field Note 11, 4 pp.

91 Canada. National Research Council. 1974. "Position report on design and certification requirements relating to bird collisions with aircraft." Assoc. Comm. on Aerospace Structure and Materials. Report No. ACASM-TN-3, 11 pp.

92 Canada. National Research Council. 1969. "Proc. World Conf. Bird Hazards to Aircraft, Kingston, Ontario, 2-5 Sept 1969." National Research Council of Canada, Ottawa, 542 pp.

93 Canada. Statistics Canada. "Canadian Civil Aviation Activity 1933-1960."

94 Canada. Transport. 1971. "Aircraft accident report P1048." Dept. Transport. Ottawa.

95 Canada. Transport. 1973. "Wildlife Control—Birds." Field Maintenance Tech. Circ. 4-13-1. Min. Transport. Ottawa. 6 pp.

96 Canada. Transport. 1972. "Standards for landscape construction of public access (class 1) areas on airports." Field Maintenance Tech. Circ. 4-1-1, Min. Transport. Ottawa. 13 pp.

97 Canada. Transport. 1972. *Flight Information Manual.* Civil Aviation Branch, Min. Transport. Ottawa. 141 pp.

98 Canada. Transport. (undated). "Bird Hazards to Aircraft." Flight Safety Pamphlet. Dept. Transport. Ottawa. n.p.

99 Canadian Air Transportation Administration. 1972. "Land use in the vicinity of airports. Planning guidelines for the use of land outside the airport property boundary." Ottawa.

100 Canadian Armed Forces. 1967. "CF104 write-offs due to bird strikes." Rept. issued by Dir. Flight Safety, Can. Forces Headquarters, Ottawa. 4 pp.

101 Canadian Armed Forces. 1967. "J79 Birdstrike Manual for investigators." Rept. issued by Dir. Flight Safety, Can. Forces Headquarters, Ottawa. 51 pp.

102 Canadian Armed Forces. 1969. "Birdstrikes 1967-68." Rept. issued by Dir. Flight Safety, Can. Forces Headquarters, Ottawa. 38 pp.

103 Canadian Armed Forces. 1972. "Bird strikes to Canadian Forces Aircraft 1969-71." Rept. issued by Dir. Flight Safety, Can. Forces Headquarters, Ottawa. 14 pp.

104 Canadian Armed Forces. 1972. "Night vision devices for the observation of birds on airfields." Rept. issued by Can. Forces Air Navigation School, CFB Winnipeg, Man. 5 pp.

105 Captain X. 1972. *Safety Last. The dangers of commercial aviation: an indictment by an airline pilot.* The Dial Press, New York.

106 Carr-Lewty, R.A. 1943. "Reactions of birds to aircraft." *British Birds* 36:151-152.

107 Catling, P.M. 1973. "Food of Snowy Owls wintering in southern Ontario with particular reference to the Snowy Owl hazard to aircraft." *Ontario field biol.* 27:41-45.

108 Clausen, P.R. 1973. "Results and description of electronic counting (intensity correlation) of birds performed in the autumn 1972 at air station Stensved." Paper for 2nd Mtg., Radar Working Group, BSCE, Brussels. 25 Jan 1973.

109 Clausen, P.R. 1973. "Results and description of electronic counting of birds, performed in the spring 1973 at air station Stensved." Paper for 8th Mtg., BSCE, Paris. May 1973.

110 Cochran, W.W., G.G. Montgomery and R.R. Graber. 1967. "Migratory flights of Hylocichla thrushes in spring: a radiotelemetry study." *Living Bird* 6:213-225.

111 Cogswell, H.L. 1969. "Gulls and solid waste disposal in the San Francisco Bay area, California." *In* 92, pp. 421-438.

112 Cogswell, H.L. 1974. "The hazard of gulls to aircraft in a bayside complex of airports and solid waste sites." *In* 161A, pp. 27-108.

113 Collum, R.O. (undated). "Operation Bahari." 401st Tactical Fighter Wing, Torrejon Air Base, Spain. 13 pp.

114 Compton, D.E. 1973. "Evaluation of the deer-aircraft collision hazard at Vandenberg AFB, California." Air Force Weapons Laboratory, Air Force Systems Command, Kirtland AFB, New Mexico, Tech. Note AFWL-DE-TN-73-022, 12 pp.

115 Compton, E., J. Boulter and E. Meyer. 1973. "Evaluation of the bird-aircraft strike hazard at Reese AFB, Texas." Air Force Weapons Laboratory, Air Force Systems Command, Kirtland AFB, New Mexico. Tech. Note AFWL-DE-TN-73-017, 16 pp.
116 Courtney, J.L. (undated). "Reduction of Herring Gull hazards, Lakehead Airport, Ontario, Canada." ACBHA, Field Note 7.
117 Cox, R.S. 1969. "Bird ingestion problems relating to gas turbine engines." *In* 92, pp. 382-389.
118 Croswell, A.R. and D.T. Horsfield. 1973. Personal communication.
119 Cuisin, M. 1963. "Rythmes d'occupation de l'aéroport de Nice-Côte d'Azur par la faune aviaire au cours d'un cycle annuel." *In* 85, pp. 77-83.
120 Deane, C.D. 1955. "Curlews at 4500 feet." *British Birds* 48:544.
121 de Boer, J. 1972. "Evaluation of bird-aircraft strike hazard (Greater Prairie Chickens) at Whiteman Air Force Base, Missouri." Air Force Weapons Laboratory, Air Force Systems Command, Kirtland AFB, New Mexico. AFWL-TN-72-025, 8 pp.
122 de Boer, J. 1974. "The United States Air Force Bird-Aircraft Strike Hazard (BASH) Program." *In* 161A, pp. 1-9.
123 de Boer, J. 1974. Personal communication.
124 de Bruin, J.P.C. and F.H.M. Mikx. 1969. "Bird Strikes. A survey of the bird problem in the Royal Netherlands Air Force, 1966-68." Mimeo. rept. issued by RNethAF. 14 pp. (in Dutch).
125 de Jong, A.P. 1970. "Their airspace or ours—a survey of progress in bird strike prevention." *Shell Aviation News* 390:2-7.
126 de Jong, A.P. and H. Blokpoel. 1966. "Hard facts about soft feathers." *Shell Aviation News* 342:2-7.
127 Denmark. Committee on Planning & Construction of Copenhagen Airport Saltholm. (undated). "Program for an off-shore Copenhagen Airport Saltholm. Summary." 17 pp.
128 Dennis, D. 1973. Personal communication.
129 Desfosses, P.P. 1967. "Roost movement observations. Metz, France." Rept. to ACBHA. 3 pp.
130 Desfosses, P.P. 1969. "Roost movement observations. Montreal International Airport 1964-1965." Rept. to ACBHA. 9 pp.
131 Desfosses, P.P. 1968. "Roost movement observations at Edmonton International Airport." Rept. to ACBHA. 5 pp.
132 Dorst, J. 1962. *The migrations of birds.* Windmill Press Ltd., Kingswood, Surrey. 476 pp.
133 Drost, R. 1960. "Uber den nächtlichen Vogelzug auf Helgoland." *In Proc. XII Internat. Ornithol. Congr.*:pp. 178-192.
134 Drury, W.H. and J.A. Kadlec. 1974. "The Herring Gull population of the East Coast: where to now?" *In* 161A, pp. 109-123.
135 Drury, W.H. and I.C.T. Nisbet. 1964. "Radar studies of orientation of song bird migrants in south-eastern New England." *Bird Banding* 35:69-119.
136 Drury, W.H. and I.C.T. Nisbet. 1969. "Strategy of management of a natural population: the Herring Gull in New England." *In* 92, pp. 441-453.
137 Duncan, C.J. 1964. "Taste". *In* 402, pp. 807-808.
138 Eastwood, E. 1967. *Radar Ornithology*. Methuen, London. 278 pp.
139 Eastwood, E., G.A. Isted and G.C. Rider. 1962. "Radar ring angels and the roosting behaviour of starlings." *Proc. Roy. Soc., B.* 156:242-267.

140 Eastwood, E. and G.C. Rider. 1965. "Some radar measurements of the altitude of bird flight." *British Birds* 58:393-426.

141 Eastwood, E. and G.C. Rider. 1966. "The grouping of nocturnal migrants." *Nature* 211:1143-1146.

142 Ennion, E.A.R. 1964. "Trapping." *In* 402, pp. 835-837.

143 Evans, W. 1916. "Birds and aeroplanes." *Scottish Naturalist* 49:39-42.

144 Fankhauser, D.P. 1966. "Movements of starlings in relation to a roost near Hanover, Pennsylvania." *Bird Banding* 37:200-203.

145 Faulkner, C.E. 1963. "Wildlife management in reverse: bird control at Boston's Logan Airport." *Pest Control* 31:26-30.

146 Finley, H.R. 1974. Personal communication.

147 Finnis, R.G. 1960. "Road casualties among birds." *Bird Study* 7:21-32.

148 Fisher, H.I. 1966. "Airplane-albatross collisions on Midway Atoll." *Condor* 68:229-242.

149 Fisher, J. 1940. *Watching Birds.* Penguin Books, Middlesex and New York, 192 pp.

150 Fisher, J. and R.T. Peterson. 1964. *The World of Birds.* Doubleday & Company, Inc., Garden City, New York. 288 pp.

151 Forsythe, D.M. 1974. "Gulls, solid waste disposal, and the bird-aircraft strike hazard." *In* 161A, pp. 17-25.

152 Fowler, H.S. and G.G. Levy. 1965. "A specific proposal for bird-impact testing on aircraft parts." ACBHA, Field Note 28, 7 pp.

153 Fowler, H.S. and G.G. Levy. 1965. "A discussion of methods of testing bird-impact on windshields and other aircraft parts." ACBHA, Field Note 26, 10 pp.

154 Frajola, W.J. 1974. "Species identification of birds involved in aircraft collisions.' *In* 161A, pp. 11-16.

155 Frings, H. and M. Frings. 1964. *Animal communication.* Blaisdell Publ. Co., New York, Toronto, London. 204 pp.

156 Frings, H. and M. Frings. 1967. "Behavioral manipulation (visual, mechanical, and acoustical)." *In* 246.

157 Gabrielson, I.N. and F.C. Lincoln. 1959. *Birds of Alaska.* Stackpole Co., Harrisburg, Pennsylvania. 927 pp.

158 Garrity, N. and P.A. Pearce. 1973. "A study of robin control in blueberry fields by use of a radio-controlled model aircraft." Rept. to Can. Wildlife Serv., Ottawa. 7 pp.

159 Gauthreaux, S.A. 1969. "A portable ceilometer technique for studying low-level nocturnal migration." *Bird Banding* 40:309-320.

160 Gauthreaux, S.A. 1970. "Weather radar quantification of bird migration." *BioScience* 20:17-20.

161 Gauthreaux, S.A. 1974. "The observation of birds with weather and airport surveillance radars." Air Force Weapons Laboratory, Air Force Systems Command, Kirtland AFB, New Mexico. Tech. Rept. AFWL-TR-74-57, 48 pp.

161A Gauthreaux, S.A. (Ed.). 1974. "Proc. Conference on the biological aspects of the bird-aircraft collision problem, Clemson University, Clemson, South Carolina, 5-7 Feb 1974." 535 pp.

162 Gehring, W. 1963. "Radar-und Feldbeobachtungen über den Verlauf des Vogelzuges im Schweizerischen Mittelland: Der Tagzug im Herbst (1957-1961)." *Ornithol. Beobach.* 60:35-68.

163 German Air Force. 1969. "Die Zugvogelbeobachtung and ihre Meldung." Besondere Anweisung für den Geophysikalischen Beratungsdienst der Bundeswehr, Nr. 105. 26 pp.

164 Glover, K.M., K.R. Hardy, T.G. Konrad, W.N. Sullivan, and A.S. Michaels. 1966. "Radar observations of insects in free flight." *Science* 154:967-972.
165 Godfrey, W.E. 1966. *The Birds of Canada*. Queen's Printer, Ottawa. 428 pp.
166 Goodhue, L.D. and F.M. Baumgartner. 1965. "The Avitrol method of bird control." *Pest Control* 33:16.
167 Graber, R.R. and W.W. Cochran. 1959. "An audiotechnique for the study of nocturnal migration of birds." *Wilson Bull.* 71:220-236.
168 Graber, R.R. and J.W. Graber. 1963. "A comparative study of bird populations in Illinois, 1906-1909 and 1956-1958." Illinois Nat. Hist. Surv. Bull. 28:377-528.
169 Graber, R.R. and S.S. Hassler. 1962. "The effectiveness of aircraft-type radar in detecting birds." *Wilson Bull.* 74:367-380.
170 Gramet, P. et J. Hanoteau. 1963. "Le problème de l'accoutumance des Corvidés et Laridés aux signeaux de détresse." *In* 85, pp. 267-272.
171 Grier, J.W. 1973. "Techniques and results of artificial insemination with Golden Eagles." *Raptor Res.* 7:1-12.
172 Griffin, D.R. 1964. *Bird migration. The biology and physics of orientation behavior*. Doubleday and Co., Inc., Garden City, New York. 180 pp.
173 Griffin, D.R. 1973. "Oriented bird migration in or between opaque cloud layers." *Proc. Amer. Phil. Soc.* 17:117-141.
174 Grudey, T.M. 1972. "Foreign body impact hazards to aircraft: an updated bibliography and discussion of possible analytical methods for the windshield bird impact problem." Nat. Aeronaut. Estab., Nat. Res. Council, Ottawa. Rept. ST-155. Revised. 40 pp.
175 Gunn, W.W.H. 1965. "The reduction of bird hazards to aircraft at Toronto International Airport." ACBHA, Field Note 34. 13 pp.
176 Gunn, W.W.H. 1966. "Radar views of migration patterns in the Great Lakes region of North America." Paper presented at 14th Internat. Ornithol. Congr., Oxford, England.
177 Gunn, W.W.H. (LGL Limited). 1972. "Bald Eagles at Port Hardy Airport." Interim Rept. No. 2. for ACBHA. 3 pp.
178 Gunn, W.W.H. (LGL Limited). (undated). "Experimental research on the use of sound to disperse Dunlin Sandpipers at Vancouver International Airport." Rept. for ACBHA. 8 pp.
179 Gunn, W.W.H. 1971. Personal communication.
180 Gunn, W.W.H. and E.P. Cockshutt. 1966. "Bird strike probability: strike predictions using observed bird density data." ACBHA, Field Note 30 (revised). 5 pp.
181 Gunn, W.W.H. and V.E.F. Solman. 1968. "A bird warning system for aircraft in flight." *In* 301, pp. 87-96.
182 Halladay, D.R. 1968. "Avian ecology as it relates to the bird hazard problem at Vancouver Airport." M.Sc. Thesis. Univ. British Columbia, Vancouver, B.C. 301 pp.
183 Hansen, C.E. 1973. "Avianological services—a success." *Flight Comment*, May-June, 1973.
184 Hanson, H.C. 1965. *The Giant Canada Goose*. Southern Illinois Univ. Press, Carbondale and Edwardsville, Illinois. 226 pp.
185 Hardenberg, J.D.F. 1963. "Clearance of birds on airfields." *In* 85, pp. 121-126.
186 Harper, W.G. 1957. " "Angels" on centimetric radars caused by birds." *Nature* 180:847-849.

187 Harper, W.G. 1958. "Detection of bird migration by centimetric radar—a cause of radar angels." *Proc. Roy. Soc., B.* 149:484-502.

188 Harper, W.G. 1959. "Roosting movements of birds and migration departures from roosts as seen by radar." *Ibis* 101:201-208.

189 Harris, R.D. 1964. "Report from Hatch in Winnipeg." ACBHA, Field Note 12, 1 pp.

190 Harris, R.D. 1964. "Bird strikes at Vancouver during hours of darkness." ACBHA, Field Note 25, 1 p.

191 Harris, R.D. and D.R.M. Hatch. 1964. "Winnipeg." ACBHA, Field Note 20, 6 pp.

192 Harris, R.D. and W.M. Hughes. 1965. "The bird strike problem at Vancouver International Airport." ACBHA, Field Note 27, 9 pp.

193 Harrison, J. 1969. "The altitude of a migrating Shoveler." *Bull. Brit. Orn. Club* 89:72.

194 Hart, P.F. 1969. "Military bird strikes in the United Kingdom." *In* 92, pp. 21-30.

195 Hartby, E. 1969. "The calls of the Starling (*Sturnus vulgaris*)." *Dansk Ornithol. Foren. Tidsskr.* 62:205-230.

196 Hassler, S.S., R.R. Graber and F.C. Bellrose. 1963. "Fall migration and weather, a radar study." *Wilson Bull.* 75:56-77.

197 Heighway, D.G. 1969. "Falconry in the Royal Navy." *In* 92, pp. 187-192.

198 Hickling, R.A.O. 1969. "The increase of gulls." *In* Gulls as pests. A discussion organized by the Seabird Group. *Ibis* 111:445.

199 Hild, J. 1968. "Beobachtungen des Kranichszuges am Niederrhein mit Hilfe von Radargeräte." *Niederrhein* 1/1968:17-21.

200 Hild, J. 1969. "Bird strikes in the German Air Force." *In* 92, pp. 31-42.

201 Hild, J. 1969. "Vögel auf Flugplätzen." Rept. issued by German Air Force, 5050 Porz-Wahn, W. Germany. 97 pp.

202 Hild, J. 1971. "Mixtures of grass-seed for airports." Paper for 6th Mtg., BSCE, Copenhagen. June 1971. 3 pp.

203 Hild, J. 1971. "Bird strikes 1970 in German Air Force." Paper for 6th Mtg., BSCE, Copenhagen. June 1971. 4 pp.

204 Hild, J. 1972. "Bird strike situation in German Air Force." Paper for 7th Mtg., BSCE, London. June 1972. 4 pp.

205 Hild, J. 1973. "Procedure of bird strike warning—forecast and advisory—in GAF." Paper for 8th Mtg., BSCE, Paris. May 1973. 2 pp.

206 Hild, J. 1974. "Das Vogelschlagproblem in der Verhütung von Störungen, Zwischenfällen und Unfällen mit Luftfahrzeugen der Bundeswehr." German Air Force, 5050 Porz-Wahn, W. Germany. Fachliche Mitteil. Nr. 176. 176 pp.

207 Hodson, N.L. 1962. "Some notes on the causes of bird road casualties." *Bird Study* 9:168-173.

208 Hooper, R.C. 1973. "Effectiveness of long-grass policy in the UK." Paper for Mtg., Airfield Bird Strike Working Group, BSCE, Munich, West Germany. Oct 1973. n.p.

209 Hoorn, A.J.W. 1973. "Research at Schiphol Airport in 1973." Paper for Mtg., Airfield Bird Strike Working Group, BSCE, Munich, West Germany. Oct 1973. 5 pp.

210 Hoorn, A.J.W. 1973. "The bird problem at Schiphol Airport, research in 1970 and 1971." Unpubl. rept. 36 pp. (in Dutch).

211 Hoorn, A.J.W. 1973. "The heron problem at Schiphol Airport." Rept. for Schiphol Airport Administration. 4 pp (in Dutch).

212 Hoorn, A.J.W. 1973. Personal communication.
213 Horeff, T.G. 1969. "Development of a prototype turbine engine inlet device for protection against bird ingestion." *In* 92, pp. 403-411.
214 Houghton, E.W. 1968. "The effect of changes in target trajectory on the wing-beat modulation pattern." Royal Radar Establishment, Malvern, UK., RRE Memo. 2456. 6 pp.
215 Houghton, E.W. 1970. "Spring migration at Gibraltar." Royal Radar Establishment, Malvern, UK., RRE Memo 2593.
216 Houghton, E.W. 1971. "ATC and bird radar surveillance without tears." Paper for 6th Mtg., BSCE. Copenhagen. June 1971. 9 pp.
217 Houghton, E.W. 1973. "Highlights of the NATO-Gibraltar bird migration radar study." Paper for 8th Mtg., BSCE, Paris. May 1973. 5 pp.
218 Hughes, W.M. 1967. "Birds trapped on Vancouver Airport banded and released. January 1964-May 15, 1967." ACBHA, Field Note 47, 20 pp.
219 Hunt, F.R. 1972. "Minutes 1st Mtg., Radar Working Group, BSCE, Malvern, England. Feb 1972."
220 Hunt, F.R. 1973. "Minutes 2nd Mtg., Radar Working Group, BSCE, Brussels. Jan 1973."
221 Hunt, F.R. 1973. "Probability of a bird strike on an aircraft." ACBHA, Field Note 62, 8 pp.
222 Hunt, F.R. 1973. "The practical aspects of microwave radiation on birds." ACBHA, Field Note 64, 9 pp.
223 Hunt, F.R. 1974. Personal communication.
224 International Civil Aviation Organization. 1969. "Bird Hazard Reduction." *In ICAO Manual.* Part 5-II-Equipment, Procedures and Service, Chapter 3. pp. 117-179.
225 ICAO Air C-WP/258, p. 8.
226 ICAO. 1972. "ICAO Airport Traffic Statistics."
227 Ireland, L.C., S.S. Ireland, T.C. Williams and J.M. Williams. 1974. "The bat hazard to USAF aircraft." *In* 161A, pp. 261-278.
228 Israel Government. 1966. "National Summary to ICAO." Attachment to state letter AN/3/31-65/190. Appendix H-Israel. Bird Strike Incident Survey Report Form. 1 July 1960 to 30 Sept. 1966.
229 Jacoby, V.E. 1969. "Bird strikes in the USSR." *In* 92, pp. 101-108.
230 Jacoby, V.E. 1970. "Some publications received from Dr. V.E. Jacoby, Moscow, USSR". ACBHA, Field Note 52, 34 pp.
231 Jacoby, V.E. 1971. "Analysis of bird strikes against aircraft." ACBHA, Field Note 55, 3 pp.
232 Jacoby, V.E. 1971. "Bird behaviour and man-made objects." ACBHA, Field Note 57, 28 pp.
233 Jacoby, V.E. 1974. *Biological basis for the prevention of bird strikes on aircraft.* "Hayka", Acad. Sci., Moscow, USSR. 167 pp. (in Russian).
234 Jacoby, V.E. and I.F. Lipinskii. 1969. "Prevention of collisions between civil aircraft and birds." Min. Civil Aviation USSR. State Res. Inst. Civil Aviation, Moscow. (in Russian).
235 Joensen, A.H. 1968. "Wildfowl counts in Denmark in November 1967 and January 1968—Methods and Results." *Danish Rev. Game Biol.*, Vol. 5, No. 5, 72 pp.
236 Joensen, A.H. 1971. "Birds and aircraft." *Dansk Vildforsk.* 1970-71:11-18. (in Danish).
237 Joensen, A.H. 1973. Personal communication.

238 Kaiser, G. 1970. "Weights of a number of birds." ACBHA, Field Note 51, 32 pp.

239 Kangas, P. and G.L. Pigman. 1949. "Development of aircraft windshields to resist impact with birds in flight. Part I. Collisions of birds with aircraft in scheduled commercial operations in the continental United States." Civil Aeronautics Admin., Indianapolis, Indiana. TDC Rept. 62.

240 Karlsson, J. 1973. "Bird research at Sturup I." Rept. for Sturup Airport Administration. PM nr F7, 1973-08-17. (in Swedish).

241 Karlsson, J. and L.-O Turesson. 1973. "Preliminary work before the opening of Malmoe/Sturup Airport for the purpose of reducing the risks of bird strikes." Paper for 8th Mtg., BSCE, Paris. May 1973. 2 pp.

242 Keil, W. 1969. "The Crane migration in Germany in connection with weather and radar." *In* 92, pp. 245-251.

243 Keil, W. 1970. "Untersuchungen über den Zug des Kranichs—Grus grus—von Herbst 1966 bis Frühjahr 1970." *Emberizia* 2:49-60.

244 Kelso, L. 1972. Review of "Some functional and morphological features of the effect of laser radiation on the pigeon retina." [E. Ter-Gazaryants and D. Gevorkyau. 1972. *Biol. Zhurn. Armenii* 25:83-87] *Bird Banding* 43:298.

245 Kenyon, K.W., D.W. Rice, C.S. Robbins, and J.W. Aldrich. 1958. "Birds and aircraft on Midway Islands, November 1956-June 1957 investigations." Fish Wildlife Serv., US Dept. Interior, Washington, D.C. Spec. Sci. Rept., Wildlife No. 38, 51 pp.

246 Kilgore, W.W. and R.L. Doutt (Eds.). 1967. *Pest Control. Biological, physical, and selected chemical methods.* Academic Press, New York. 477 pp.

247 KLM Royal Dutch Airlines. 1968. "Survey of bird strikes. 1963 through 1967." Rept. issued by Flight Operations Division, Flight Safety Department, KLM. n.p.

248 Kryter, K.D. 1970. *The effects of noise on man.* Academic Press, New York, London. 633 pp.

249 Kuhring, M.S. and T. Hammell. 1963. "Distress calls from Pigeons." ACBHA, Field Note 1, 3 pp.

250 Kuhring, M.S. 1974. Personal communication.

251 Lack, D. 1960. "The height of bird migration." *British Birds* 53:5-10.

252 Lack, D. 1963. "Migration across the southern North Sea studied by radar. Part 4. Autumn." *Ibis* 105:1-54.

253 Laham, Q.N. 1967. "Report on aircraft turbine engine birdstrike investigations." ACBHA, Field Note 43, 27 pp.

254 Laham, Q.N. 1969. "Amino acid analysis and bird identification." *In* 92, pp. 415-417.

255 Langley, M. 1970. "The birdstrike problem." *Science J.* 6:55-60.

256 Laskey, A.R. 1972. "One-day or one-night mileage of migrant birds." *Bird Banding* 43:287.

257 Laty, M. 1965. "Local movements of Laridae in the region of Aix-en-Provence, France." Mimeo. Rept.

258 Laty, M. 1970. "Localization of collision risk. Birds-Aircraft." Rept. issued by General Secretariat for Civil Aviation, France.

259 Laty, M. 1972. "Détection par le radar MELODI d'oiseaux en migration dans la région des Treppes." *Météorol.* Jan-March 1972, pp. 79-91.

260 Laty, M. 1973. "Le problème "oiseaux" en France: activité de recherche." Paper for 8th Mtg., BSCE, Paris. May 1973.

261 Laty, M. 1973. Personal communication.

262 Lavery, H.J. 1969. "Collisions between aircraft and birds, at Townsville, *Queensland. J. Agr. Anim. Sci.* 26:447-455.
263 Lebret, T. 1970. "Noctural feeding and other activities of the Grey Lag Goose (Anser anser) in freshwater tidal habitat in the Netherlands." *Limosa* 43:11-30. (in Dutch with English summary).
264 Lewis, H.F. 1970. ACBHA, Bulletin 5, 34 pp.
265 LGL Ltd. 1974. "The bird hazard problem at Vancouver International Airport. Autumn 1973." Rept. for the National Research Council of Canada, 101 pp.
266 LGL Ltd. 1974. "A series of proposals for further research on the bird-hazard problem at Vancouver International Airport," Rept. to ACBHA, National Research Council of Canada.
267 LGL Ltd. 1974. "Executive report on bird populations and movements associated with the proposed site of the new Toronto International Airport." Rept. for General Manager, Toronto Area Airports Project, Ministry of Transport, Canada. 100 pp.
268 Lid, G. 1973. "The bird strike problem in Norway." Paper for 8th Mtg., BSCE, Paris. May 1973. 11 pp.
269 Lid, G. 1973, Personal communication.
270 Lincoln, F.C. 1950. *Migration of Birds.* Fish Wildlife Serv., US Dept. Interior, Washington, D.C. Circular 16, 102 pp.
271 Lind, H. 1971. "An attempt to reduce the Herring Gull population at Saltholm." Paper for 6th Mtg., BSCE, Copenhagen. June 1971.
272 Lind, H. 1973. "Bird strike problems on civil airports in Denmark and our treatment of those problems." Paper for Mtg., Aerodrome Bird Strike Working Group, BSCE, Munich, West Germany. Oct 1973.
273 Louette, M. 1971. "Radar data on the fall migration in 1969 and 1970 across northwestern Belgium. (Direction of migration)." *De Wielewaal,* Dec 1971, pp. 345-350. (in Dutch).
274 Louette, M. 1971. "Différence d'intensité de migration entre la zone cotière Belge et l'intérieur du pays vue par radar." *Bull. Aves* 8:41-45.
275 Louette, M. 1973. "The cancellation of the 8-point scale measurement for migration studies in Belgium." Paper for 2nd Mtg., Radar Working Group, BSCE, Brussels. Jan 1973.
276 Low, J.B., L. Kay and D.I. Rasmussen. 1950. "White Pelicans on Great Salt Lake." *Auk* 67:345-356.
277 Lowery, G.H. and R.J. Newman. 1955. "Direct studies of nocturnal migration." *In* A. Wolfson (Ed.). Recent studies in avian biology. Univ. Illinois Press, pp. 238-263.
278 Lowery, G.H. and R.J. Newman. 1966. "A continentwide view of bird migration on four nights in October." *Auk* 83:547-586.
279 Lustick, S.I. 1972. "Physical techniques for controlling birds to reduce aircraft strike hazards (effects of laser light on bird behavior and physiology)." Air Force Weapons Laboratory, Air Force Systems Command, Kirtland AFB, New Mexico. Tech. Rept. AFWL-TR-72-159, 46 pp.
280 Lustick, S.I. 1973. "The effect of intense light on bird behavior and physiology." Proc. 6th Bird Control Seminar, Bowling Green State Univ., Bowling Green, Ohio, 30 Oct-1 Nov 1973. pp. 171-186.
281 MacGibbon, M.R. 1973. "Model plane trials." Minutes 60th Mtg., ACBHA, Nat. Res. Council Can., Ottawa.
282 Macauley, G.A. 1965. "Some thoughts on the forces associated with a bird strike." ACBHA, Field Note 24, 3 pp.

283 Maldague, M. et al. 1973. "Research on etho-ecological methods for eliminating earthworms from the vicinity of airport runways." Rept. to ACBHA, Nat. Res. Council Can., Ottawa. 109 pp.
284 Manville, R.H. 1963. "Altitude record for Mallard." *Wilson Bull.* 75:92.
285 Matthews, G.V.T. 1964. "Navigation." *In* 402, pp. 510-513.
286 Mattingly, E.B., O. Collum and J. de Boer. 1973. "Falconry as a means of reducing bird-aircraft strike hazards at Whiteman Air Force Base, Missouri. A progress report." Kirtland AFB, New Mexico, Tech. Note DE-TN-73-007. 11 pp.
287 McNeil, R. 1972. "Contrôle des populations d'oiseaux pour la sécurité aérienne. Rap. prélim. No. 2." Centre Recherches Ecol. de Montréal. 219 pp.
288 Mead, C. 1973. "How they turn night into day." *BTO News* No. 58. May 1973.
289 Meinertzhagen, R. 1955. "The speed and altitude of bird flight (with notes on other animals)." *Ibis* 97:81-117.
290 Meng, H. 1971. "The Swedish Goshawk Trap." *J. Wildlife Mgmt.* 35:832-845.
291 Miller, A.H. 1957. "Migratory flight of a Zonotrichia at 10,000 feet above ground level." *Condor* 59:209-210.
292 Mitchell, J. 1966. "More thoughts on forces associated with a bird strike." ACBHA, Field Note 37, 7 pp.
293 Mitchell, K.D.G. 1955. "Aircraft observations of birds in flight." *British Birds* 48:59-70.
294 Mitchell, K.D.G. 1957. "Further aircraft observations of birds in flight." *British Birds* 50:291-302.
295 Mitchell, K.D.G. 1964. "Further observations of birds from aircraft." *British Birds* 57:315-324.
296 Mitchell, R.T. 1963. "The floodlight trap, a device for capturing large numbers of blackbirds and starlings at roosts." Wildlife Res., Patuxent Wildlife Res. Center., US Dept. Interior, Spec. Sci. Rept., Wildlife no. 77. 13 pp.
297 Moreau, R.E. 1953. "Migration in the Mediterranean area." *Ibis* 95:329-364.
298 Moreau, R.E. 1972. *The Palaearctic-African Bird Migration Systems.* Academic Press, London, New York. 384 pp.
299 Mörzer Bruyns, M.R., J. Phillipona and A. Timmerman. 1969. "Survey of the winter distribution of Palaearctic geese in Europe, Western Asia and North Africa." Rept. issued by the Goose Working Group of the International Waterfowl Research Bureau. Zeist, The Netherlands. 111 pp.
300 Munro, W.T. 1973. Personal communication.
301 Murton, R.K. and E.N. Wright (Eds.). 1968. *The Problems of Birds as Pests.* Institute of Biology Symposia No. 17. Academic Press. London and New York, 254 pp.
302 Myres, M.T. 1964. "Technical details of radar equipment detecting birds, and a bibliography of papers reporting the observation of birds with radar." ACBHA, Field Note 9, 20 pp.
303 Myres, M.T. 1969. "The detection of birds, and study of bird movements, with radar." *In* 92, pp. 501-519.
304 Myres, M.T. 1969. Comments during discussion. *In* 92, pp. 193.
305 Myres, M.T. and S.R. Canning. 1971. "A Canada Goose migration through the southern interior of British Columbia." *In* Studies of bird hazards to aircraft. Can. Wildlife Serv. Rept. Ser. No. 14:23-34.
306 Neff, J.A. 1964. "Banded Evening Grosbeak strikes airplane." *Bird Banding* 35:265.

307 Nelson, A.L. and A.C. Martin. 1953. "Game bird weights." *J. Wildlife Mgmt.* 17:36-41.

308 Nelson, S.O. and J.L. Seubert. 1966. "Electromagnetic and sonic energy for pest control." *In Scientific aspects of pest control.* Nat. Acad. Sci., Nat. Res. Council, Washington, D.C. Publ. 1402. pp. 135-156.

309 Nemergut, J.P. and R.C. Beason. 1972. "Preliminary evaluation of bird-aircraft strike hazard at Beale AFB, California." Air Force Weapons Laboratory, Air Force Systems Command, Kirtland AFB, New Mexico. Consultation Rept. 5, DE-TN-72-018, 16 pp.

310 Nemergut, J.P. 1973. "A review of the USAF bird strike problem." *In* 414, pp. 425-445.

311 Newman, R.J. and G.H. Lowrey. 1964. "Selected quantitative data on night migration in autumn." Louisiana State Univ., Special Publs. Mus. Zool. 3:1-39.

312 New Zealand Ministry of Transport. 1971. "New Zealand Civil Aircraft Accidents." Office of Air Accidents Investigation, Ministry of Transport, Wellington.

313 Nisbet, I.C.T. 1963. "Quantitative study of migration with 23-centimetre radar." *Ibis* 105:435-460.

314 Nisbet, I.C.T. 1963. "Measurements with radar of the height of nocturnal migration over Cape Cod, Massachusetts." *Bird Banding* 34:57-67.

315 Nisbet, I.C.T. 1969. Comments during discussion. *In* 92, p. 193.

316 Nisbet, I.C.T. and W.H. Drury. 1967. "Scanning the sky: birds on radar." *Massachusetts Audubon*, summer 1967.

317 Noer, H. 1971. "Recent developments of the Danish bird migration forecast system." Paper for 6th Mtg., BSCE, Copenhagen. June 1971.

318 Noer, H., J. Rabøl and A.H. Joensen. 1970. "A forecast model for bird migration in Denmark." Prelim. rept. 14 pp.

319 Orians, G.H. and G.M. Christman. 1968. *A comparative study of the behavior of Red-winged, Tricolored, and Yellow-headed Blackbirds.* Univ. Calif. Publ. Zool., Vol. 84. Univ. Calif. Press, Berkeley and Los Angeles. 81 pp.

320 Oranje, J. and P. van der Wielen. 1968. "Warnings of and research on bird migration using radar". Rept. issued by the Scientific Research Section, RNethAF, The Hague. 20 pp. (in Dutch).

321 Peake, D.J. 1967. "The resistance of aircraft windshields to bird impact." Quarterly Bull. Div. Mech. Eng. and Nat. Aeronaut. Estab., Nat. Res. Council Can., Ottawa, Rept. DME/NAE 1967(3). 43 pp.

322 Pennycuick, C.J. 1969. "The mechanics of bird migration." *Ibis* 111:525-556.

323 Pennycuick, C.J. 1972. "Soaring behaviour and performance of some East African birds, observed from a motor glider." *Ibis* 144:178-218.

324 Perrins, C.M. 1964. "Weights." *In* 402, pp. 883-884.

325 Pettingill, O.S. 1970. *Ornithology in laboratory and field.* Burgess Publ. Co., Minneapolis, Minn., 4th ed. 524 pp.

326 Porter, R. and D.S.M. Wiemeyer. 1970. "Propagation of captive American Kestrels." *J. Wildlife Mgmt.* 34:594-604.

327 Pumphrey, R.J. 1964. "Hearing and balance." *In* 402, pp. 358-359.

328 Rice, D.W. 1959. "Birds and aircraft on Midway Islands, 1957-58 investigations." Fish Wildlife Serv., US Dept. Interior, Washington, D.C. Spec. Sci. Rept., Wildlife No. 44. 49 pp.

329 Rice, D.W. and K.W. Kenyon. 1962. "Breeding distribution, history and populations of North Pacific albatrosses." *Auk* 79:365-386.

330 Richardson, W.J. 1969. "Temporal variations in the volume of bird migration: a radar study in Canada." *In* 92, pp. 323-334.

331 Richardson, W.J. 1972. "Temporal variations in the ability of individual radars in detecting birds." ACBHA, Field Note 61, 57 pp.

332 Richardson, W.J. 1972. "Autumn migration and weather in eastern Canada: a radar study." *American Birds* 26:10-17.

333 Richardson, W.J. 1974. "Multivariate approaches to forecasting day-to-day variations in the amount of bird migration." *In* 161A, pp. 309-329.

334 Richardson, W.J. 1974. "Spring migration over Puerto Rico and the western Atlantic, a radar study." *Ibis* 166:172-193.

335 Richardson. W.J. and M.E. Haight. 1970. "Migration departures from starling roosts." *Can. J. Zool.* 48:31-39.

336 Ridenour, L.N. (Ed.). 1947. *Radar Systems Engineering.* Radiation Laboratory Series 1. McGraw-Hill, New York and London. 748 pp.

337 Robbins, C.S. 1945. "Live trapping of hawks and owls." *J. Wildlife Mgmt.* 9, no. 2.

338 Robin-Gunn Ltd. 1972. "The bird strike problem." A series of five illustrated lectures prepared for ACBHA, Nat. Res. Council Can., Ottawa. 81 pp.

339 Rodrigues de la Fuente, F. 1971. "Falconry for the control of birds dangerous on airports. Results of three years of practice." Paper for 6th Mtg., BSCE, Copenhagen. June 1971. 3 pp.

340 Romer, J.D. 1969. "Bird Problems at Hong Kong Airport." *In* 92, pp. 77-86.

341 Royal Air Force. 1972. "Bird concentration areas." Directorate of Flight Safety. n.p.

342 Royal Netherlands Air Force. 1969. "Goshawks at Leeuwarden Air Base." Rept. issued by Flight and Ground Safety Section, RNethAF, The Hague. 20 pp.

343 Royal Netherlands Air Force. 1974. "Bird strike analysis, military aircraft, Royal Netherlands Air Force, 1973." Rept. issued by Flight and Ground Safety Section, RNethAF, The Hague. 9 pp.

344 Samolewicz, J.J. 1964. "Temperature preference test of seagulls." Nat. Res. Council Can., Ottawa. Test Report MET-460. 8 pp.

345 Saul, E.K. 1967. "Birds and aircraft: a problem at Auckland's new international airport." *J. Roy. Aeronaut. Soc.* 7:366-376.

346 Schafer, E.W. 1970. "Problems in developing new chemicals for bird control." Proc. 5th Bird Control Seminar, Bowling Green State Univ., Bowling Green, Ohio. Sept 1970. pp. 7-10.

347 Schaefer, G.W. 1968. "Bird recognition by radar: a study in quantitative radar ornithology." *In* 301, pp. 53-86.

348 Schaefer, G.W. 1968. Remarks during discussion. *In* 301, p. 50.

349 Schaefer, G.W. 1969. "Bird hazard to aircraft: its magnitude, major causes, and reduction by radar detection." Mimeo. rept. distributed at World Conf. Bird Hazards to Aircraft, Kingston, Ontario, 2-5 Sept 1969. 11 pp.

350 Schaefer, G.W. 1969. "An airport bird detection radar for reducing bird hazards to aircraft." Rept. issued by ACBHA. 30 pp.

351 Schneider, E.P. 1973. "The result of the preventive bird-strike work in Denmark." Paper for 8th Mtg., BSCE, Paris. May 1973. 2 pp.

352 Schubert, H.J. 1957. "Uber Begegnungen von Vögeln mit Flugzeugen." *Beitr. z. Vogelkunde* 5:188-200.

353 Schüz, E. 1971. *Grundriss der Vogelzugskunde.* Verlag Paul Parey, Berlin and Hamburg. 390 pp.

354 Scott, A.G. 1964. "Falconry." *In* 402, pp. 269-271.

355 Seaman, E.A. Personal communication to M.S. Kuhring.
356 Seubert, J.L. 1963. "Biological studies of the problem of bird hazard to aircraft." *In* 85, pp. 143-168.
357 Seubert, J.L. 1965. "Biological studies of the problem of bird hazard to aircraft." FAA/BRD A-90, Rept. RD-65-73, US Dept. Interior, Washington, D.C. 22 pp.
358 Seubert, J.L. and B. Meanley. 1974. "Relationship of blackbird/starling roosts to bird hazards at airports." *In* 161A, pp. 209-220.
359 Skolnik, M.I. 1962. *Introduction to radar systems.* McGraw-Hill, New York—San Francisco—Toronto—London, 648 pp.
360 Sladen, W.J.L., W.W.H. Gunn and W.W. Cochran. 1969. "Studies on the migrations of the Whistling Swan, 1969." *In* 92, pp. 231-244.
361 Sladen, W.J.L. and W.W. Cochran. 1969. "Studies of the Whistling Swan, 1967-1968." Trans. 34th North Amer. Wildlife and Natural Res. Conf.:42-50.
362 Slot, J.W. and F.H.M. Mikx. 1968. "Report on the experimental project with Goshawks at Leeuwarden Air Base." Rept. issued by Flight and Ground Safety Section, RNethAF, The Hague. 10 pp (in Dutch).
363 Smith, R.N. 1966. "Control of birds in aircraft hangars." Proc. 3rd. Bird Control Seminar, Bowling Green State Univ., Bowling Green, Ohio. pp. 30-33.
364 Sobieralski, R.J. 1972. "Ecological aspects of bird-aircraft strike hazards." Air Force Weapons Laboratory, Air Force Systems Command, Kirtland AFB, New Mexico. Tech. Rept. AFWL-TR-71-94, 24 pp.
365 Solman, V.E.F. 1965. "Use of falcons for airport bird control." ACBHA, Field Note 33, 4 pp.
366 Solman, V.E.F. 1969. Rapporteur's report. *In* 92, p. 100.
367 Solman, V.E.F. 1969. "Photography in bird control for air safety." J. Biol. Photog. Assoc. 37:150-155.
368 Solman, V.E.F. 1971. "Report from the bird/radar/weather group." Paper for 6th Mtg., BSCE, Copenhagen. June 1971.
369 Solman, V.E.F. 1971. "How we reduce bird hazards to aircraft." Tenth National Conference on Environmental Effects on Aircraft and Propulsion Systems. Trenton, New Jersey, USA. 19 May 1971.
370 Solman, V.E.F. 1973. "Brief report on two major bird strikes." Rept. for 8th Mgt., BSCE, Paris. May 1973.
371 Solman, V.E.F. 1973. "Birds positively identified as involved in collisions with aircraft in Canada." List distributed to ACBHA. 2 pp.
372 Solman, V.E.F. 1973. "A collection of papers presented by the Chairman during 1973." ACBHA, Field Note 66, 40 pp.
373 Solman, V.E.F. 1973. "Aircraft and wildlife." Paper for Symposium on Wildlife in an Urbanizing Environment. Springfield, Massachusetts, USA. 27-29 Nov 1973.
374 Solman, V.E.F. 1974. Personal communication.
375 Solman, V.E.F. undated. "A proposed program for Belgium, Denmark, France, Germany, Holland, Norway, and the United Kingdom for forecasting bird migration hazardous to aircraft." Rept. for BSCE.
376 Speirs, J.M., J.J.C. Kanitz and J. Novak. 1971. "Numbers, speeds, and directions of migrating geese from analysis of a radar display at Fort William, Ontario." *In* Studies of bird hazards to aircraft. Can. Wildlife Serv. Rept. Ser. No. 14:69-76.
377 Steidinger, P. 1972. "Der Einfluss des Windes auf die Richtung des nächtlichen Vogelzuges." *Ornithol. Beobacht.* 69:20-39.

378 Stephen, W.J.D. 1961. "Experimental use of acetylene exploders to control duck damage." Trans. 26th North Amer. Wildlife and Natural Res. Conf.:98-110.

379 Stephen, W.J.D. 1967. "Bionomics of the Sandhill Crane." Can. Wildlife Serv. Rept. Ser. No. 2, 40 pp.

380 Stephen, W.J.D. and J. Ahrens. (undated). "Birds and Vancouver Airport." Rept. for ACBHA. 24 pp.

381 Stewart, R.E. and H.A. Kantrud. 1972. "Population estimates of breeding birds in North Dakota." *Auk* 89:766-788.

382 Stortenbeker, C.W. 1967. "Research on the background of the bird problem at Schiphol Airport." Mimeo. rept. 17 pp. (in Dutch).

383 Stortenbeker, C.W. 1969. "Ground cover at Schiphol Airport." *In* 92, pp. 161-166.

384 Stout, F., J.L. Hayward and H. Gillett. 1974. "Distribution of aggregations of the Glaucous-winged Gull at Shemya AFB, Alaska, and methods for their dispersal." *In* 161A, pp. 125-148.

385 Sugg, R.W. 1963. Comments during discussion. *In* 85, p. 69.

386 Sugg, R.W. 1965. "An investigation into bird densities which might be encountered by an aircraft during take-off and landing." Tech. Inf. Library Services, UK Min. Aviation, S. & T. Memo.6/65.

387 Swedish Board of Civil Aviation. 1973. Notam A 1179/73, 20 Sept 1973. Aeronautical Information Service, Bromma, Sweden. 1 p.

388 Symonds, A.E.J. 1961. "The counting of starlings at country roosts." *Bird Study* 8:185-193.

389 Taber, R.D. and I. McTaggart-Cowan. 1966. "Capturing and marking wild animals." *In Wildlife Management Techniques*. Wildlife Society, Washington, D.C. pp. 277-317.

390 Tanner, J.A. 1965. "The effect of microwave radiation on birds. Some observations and experiments." ACBHA, Field Note 31, 4 pp.

391 Tanner, J.A. 1966. 'Effect of microwave radiation on birds." *Nature* 210:636.

392 Tanner, J.A., S.J. Davie, C. Romero-Sierra and F. Villa. 1969. "Microwaves—A potential solution to the bird hazard problem in aviation." *In* 92, pp. 215-221.

393 Tanner, J.A., C. Romero-Sierra and S.J. Davie. 1967. "Non-thermal effects of microwave radiation on birds." *Nature* 216:1139.

394 Tansley, K. 1964. "Vision." *In* 402, pp. 859-865.

395 Temple, S.A. 1972. "Artificial insemination with imprinted birds of prey." *Nature* 237:287-288.

396 Tengeler, I. 1972. "An electronic counting and registration system of bird echoes with aid of radar and the possibility of processing with an international intensity scale." Lab. for Electronic Developments of the Armed Forces, Oegstgeest, Holland. Rept. 16.147/0.3271. 9 pp.

397 Tengeler, I. 1973. "Detection of birds by radar." Lab. for Electronic Developments of the Armed Forces, Oegstgeest, Holland. Rept. LEOK No. 18. 705/o.3271.

398 Thiessen, G.J., E.A.G. Shaw, R.D. Harris, J.B. Gollop and H.R. Webster. 1957. "Acoustic irritation threshold of Peking Ducks and other domestic and wild fowl." *J. Acoustical Soc. Am.* 29:1301-1306.

399 Thiessen, G.J. and E.A.G. Shaw. 1959. "Acoustic irritation threshold of Ringbilled Gulls." *J. Acoustical Soc. Am.* 29:1307-1309.

400 Thomas, G.J. 1972. "Review of gull damage and management methods at nature reserves." *Biol. Conservation* 4:117-127.

401 Thomson, A.L. 1964. "Migration." *In* 402, pp. 465-472.

402 Thomson, A.L. 1964. *A new dictionary of birds*. T. Nelson Ltd., London. 928 pp.

403 Thomson, A.L. 1968. Remarks during discussion. *In* 301, p. 39.

404 Thorpe, W.H. 1961. *Bird song: the biology of vocal communication and expression in birds.* Cambridge.

405 Thorpe, W.H. 1964. "Singing." *In* 402, pp. 739-750.

406 Thorpe, J. 1973. "Bird strikes to United Kingdom civil aircraft 1966-1971." UK Civil Aviation Authority, Airworthiness Tech. Note No. 106, Issue 1, 40 pp.

407 Thorpe, J. 1973. Personal communication. (Data compiled from files of Civil Aviation Authority, Redhill, Surrey, UK).

408 Tinbergen, N. 1964. "Releaser." *In* 402, pp. 687-688.

409 Tinbergen, L. 1967. *Birds en route*. Thieme en Cie, Zutphen, Holland. 128 pp. (in Dutch).

410 Tudor, A.J. 1969. "Experimental techniques in bird ingestion research." *In* 92, pp. 393-401.

411 Tureson, L.O. 1972. "Working program aimed at reduction of risks of bird strikes in the Malmoe/Sturup Airport area." Mimeo rept. 3 pp.

412 UK Dept. Trade and Industry. 1971. "Bird control on aerodromes." Aerodromes Tech. Dir., DAT Paper 14, Aug 1971, 39 pp.

413 US Air Force. 1967. "Natural Resources—Bird Control." Air Training Command, Randolph AFB, Texas. ATC Pamphlet 126-1, 9 pp.

414 US Air Force Systems Command. 1973. "Proc. Conference on Transparent Aircraft Enclosures, Las Vegas, Nevada, 5-8 Feb 1973." (Compiled by R.E. Whittman). Air Force Materials Laboratory, Wright-Patterson AFB, Ohio. Tech. Report AFML-TR-73-126.

415 US Air Force, Directorate of Aerospace Safety. 1974. "USAF bird strikes summary 1 Jan-31 Dec 1973." US Air Force Inspection and Safety Center, Norton AFB, California.

416 US Department of Defence. 1973. *Flight Information Publication, Section II, Planning data and procedures.* Issued by Defence Mapping Agency Aero Space Center, St. Louis Air Force Station, Missouri. 256 pp.

417 US Department of Transport. 1972. "Bird Hazards to Aircraft." Fed. Aviation Admin., Advisory Circ. 150/5200-3A, 7 pp.

418 US National Academy of Sciences. 1970. *Principles of plant and animal pest control. Vol. 5. Vertebrate problems and controls.* Washington, D.C. 153 pp.

419 US National Transportation Safety Board. 1973. "Aircraft Accident Report, Machinery Buyers Corporation, Learjet Model 24, N454RN, Atlanta, Georgia. Feb. 26, 1973." National Transportation Safety Board, Washington, D.C. Rept. NTSB-AAR-73-12, 10 pp.

420 US National Transportation Safety Board. 1973. "Aircraft Accident Report. Northwest Airlines, Inc. Boeing 747-151, N602US, Miami, Florida." National Transportation Safety Board, Washington, D.C. Rept. NTSB-AAR-73-13, 9 pp.

421 van der Heyde, J.M.M. 1963. "Moyens utilisés pour éloigner les mouettes et autres oiseaux de l'aéroport de Schiphol." *In* 85, pp. 127-130.

422 van der Heyde, J.M.M. 1973. Personal communication.

423 van Tets, G.F. 1965. "Australian airport bird problem investigations." Rept. issued by Div. Wildlife Res., CSIRO, Canberra. 9 pp.

424 van Tets, G.F. 1966. "A photographic method of estimating densities of bird flocks in flight." *CSIRO Wildlife Res.* 11:103-110.

425 van Tets, G.R. 1969. "Diurnal movement patterns of the Silver Gull, *Larus Novae-hollandiae* Stephens, at Sydney Airport. 1963-64." *CSIRO Wildlife Res.* 14:111-116.

426 van Tets, G.F. 1969. "Quantitative and qualitative changes in habitat and avifauna at Sydney Airport. Jan. 64-June 68." *CSIRO Wildlife Res.* 14:117-128.

427 van Tets, G.F. and W.J.M. Vestjens. 1973. "Birds and their food at Mackay Airport, Queensland." CSIRO, Canberra, Tech. Memo. 8. 71 pp.

428 van Tets, G.F., W.J.M. Vestjens and E. Slater. 1969. "Orange runway lighting may help reduce the bird hazard to aircraft on airfields." *In* 92, pp. 183-185.

429 van Tyne, J. and A.J. Berger. 1966. *Fundamentals of ornithology*. J. Wiley and Sons, New York. 624 pp.

430 Vermeer, K. and B. Switzer. 1968. "Road kills of birds and mammals in southeastern Alberta." *Blue Jay* 26:93-94.

431 Voous, K.H. 1959. "High-altitude flight of a migrating duck." *Limosa* 32:70. (in Dutch).

432 Weishaupt, M.G. 1969. "Surveillance radar des déplacements d'oiseaux en France. Stages de formation des controleurs radar." *In* 92, pp. 487-490.

433 Weseloh, C. (in prep.). "Local distribution patterns of gulls in relation to their seasonal centers of activity and to urban garbage dumps at Calgary, Alberta." PhD. Thesis, Univ. of Calgary, Alberta.

434 Williams, T.C. and J.M. Williams. 1969. "An investigation of the collisions of bats and birds with high performance aircraft." US Air Force Office of Scientific Research, Office of Aerospace Research, Arlington, Virginia. Tech. Rept. AFOSR 69-1021TR, 31 pp.

435 Williams, T.C. and M. Williams. 1969. "Bat hazards to aircraft: detection by radar." *In* 92, pp. 307-319.

436 Williams, J.M., T.C. Williams and L.C. Ireland. 1974. "Bird migration over the North Atlantic." *In* 161A, pp. 359-382.

437 Wiltschko, W. 1972. "The influence of magnetic total intensity and inclination on directions preferred by migrating European Robins (*Erithacus rubecula*)." Proc. Symposium NASA SP-262 Animal Orientation and Navigation. pp. 569-578.

438 Wiltschko, W. and R. Wiltschko. 1972. "Magnetic compass of European Robins." *Science* 176:62-64.

439 Wing, L.W. 1956. *Natural history of birds: a guide to ornithology*. Ronald Press Co., New York.

440 Winmill, A.E. 1969. "Chemical control of bird attractants (worms) at Canadian Forces Base Trenton." *In* 92, pp. 167-179.

441 Woodford, M.H. 1966. *A Manual of Falconry* (2nd edition). Branford. 194 pp.

442 Wooten, R.C., M.J. Boulter and W.M. Floyd. 1973. "Avian hazards to the Air Force Rocket Sled Test Track, Holloman AFB, New Mexico." Air Force Weapons Laboratory, Air Force Systems Command, Kirtland AFB, New Mexico. Tech. Rept. AFWL-TR-72-241, 22 pp.

443 Wright, E.N. 1963. "A review of bird scaring devices." *In* 85, pp. 113-119.

444. Wright, E.N. 1967. "Control of bird damage on modern airports." *Internat. Pest Control* 9:29-33.

445 Wright, E.N. 1969. "Bird dispersal techniques and their use in Britain." *In* 92, pp. 207-214.

Photo Credits

The author and publisher wish to thank the following individuals and organizations for the illustrations listed below.

front cover: Ontario Ministry of Industry and Tourism

p. 4 Canadian Wildlife Service, photo by C. Duhaime
p. 30 Canadian Armed Forces
p. 35 Air Canada
p. 39 Canadian Armed Forces (both)
p. 40 Royal Air Force
p. 41 Canadian Armed Forces (all)
p. 42 Canadian Armed Forces (both)
p. 43 Royal Air Force (both)
p. 44 Air Canada
p. 61 Royal Danish Air Force
p. 68 National Research Council, Ottawa
p. 85 United States Air Force
p. 87 National Research Council, Ottawa (all)
p. 90 National Research Council, Ottawa, photo by J.A. Tanner
p. 98 *The Globe and Mail,* Toronto
p. 105 Royal Netherlands Air Force
p. 110 H. Blokpoel
p. 116 (top left) German Air Force
(top right and bottom) H. Blokpoel
p. 119 Royal Netherlands Air Force
p. 122 National Research Council, Ottawa
p. 156 Canadian Wildlife Service, photo by H. Blokpoel
p. 184 United Press International

back cover: United Press International

Every effort has been made to acknowledge correctly the source of illustrations reproduced in this book. The publisher welcomes any information which will enable him to rectify, in subsequent editions, any errors or omissions which may have been made in giving a credit line.

Index

Numbers indicate pages where the indexed subject is introduced in the text or where additional information is given in the text. See References for names of authors.